SCIENCE IN A COMPETITIVE AND GLOBALIZING WORLD

SCIENCE POLICIES AND PROGRAMS

HISTORY, FUNDING AND ISSUES

SCIENCE IN A COMPETITIVE AND GLOBALIZING WORLD

Additional books and e-books in this series can be found on Nova's website under the Series tab.

SCIENCE IN A COMPETITIVE AND GLOBALIZING WORLD

SCIENCE POLICIES AND PROGRAMS

HISTORY, FUNDING AND ISSUES

JOHNNIE RODGERS

NOTICE TO THE READER

Library of Congress Cataloging-in-Publication Data

ISBN: 978-1-53614-107-8

Published by Nova Science Publishers, Inc. † New York

CONTENTS

PREFACE

The first chapter of this book is a fact sheet, providing data on past, current and proposed NASA appropriations. The second chapter provides an overview of the portion of Department of Defense (DOD) research, development testing and evaluation (RDT&E) funding referred to as Defense Science and Technology (Defense S&T). It provides perspectives on the role of Defense S&T in supporting US defense capabilities, historical funding levels, recent funding trends and approaches to determining how much the federal government should invest in Defense S&T, particularly in basic research. The next chapters focus on the appropriations and funding history of the National Science Foundation (NSF), which supports basic research and education in the non-medical sciences and engineering. NSF is a major source of federal support for US university research, especially in certain fields such as computer science. It is also responsible for significant shares of the federal science, technology, engineering and mathematics (STEM) education program portfolio and federal STEM student aid and support. The next chapter provides an overview of the history of science and technology (S&T) advise to the President and discusses selected recurrent issues for Congress regarding the Office of Science and Technology Policy (OSTP) and their management and operations. Finally, Science and technology (S&T) have a pervasive influence over a wide range of issues confronting the nation.

The last chapter of this book briefly outlines an array of science and technology policy issues that may come before the 115th Congress, including but not limited to agriculture, defense, energy, homeland security, information technology, physical and material sciences and space.

In: Science Policies and Programs
Editor: Johnnie Rodgers
ISBN: 978-1-53614-107-8

Chapter 1

NASA Appropriations and Authorizations: A Fact Sheet*

Daniel Morgan

Congressional deliberations about the National Aeronautics and Space Administration (NASA) often focus on the availability of funding. This fact sheet provides data on past, current, and proposed NASA appropriations. No bills have yet been introduced in the 115th Congress proposing future-year authorizations of NASA appropriations.

Table 1 shows appropriations for NASA from FY2013 through FY2018. The data for FY2013 through FY2016 include supplemental appropriations, rescissions, transfers, reprogramming, and, in the case of

* This is an edited, reformatted, and augmented version of a Congressional Research Service report, R43419, dated April 16, 2018.

FY2013, sequestration. They are taken from NASA's congressional budget justifications for FY2014 through FY2018.[1] Congressional budget justifications are available on the NASA budget website (http://www.nasa.gov/news/budget/) for the current year and for past years back to FY2002. The table data for FY2017 are as enacted by the Consolidated Appropriations Act, 2017 (P.L. 115-31). For amounts not specified in that act, see pp. H3374- H3375 of the explanatory statement, published in the *Congressional Record* on May 3, 2017. The table data for FY2018 are as enacted by the Consolidated Appropriations Act, 2018 (P.L. 115- 141). For amounts not specified in that act, see pp. H2094-H2096 of the explanatory statement, published in the *Congressional Record* on March 22, 2018.

The Administration's budget request for FY2019 adopted a new account structure for NASA. It presented FY2017 amounts adjusted for comparability with the new structure, but not FY2018 amounts, because final FY2018 appropriations had not yet been enacted at the time the FY2019 budget was released. Table 2 shows FY2017 appropriations, adjusted for comparability, and the Administration's request for FY2019. Additional columns will be added to this table as Congress acts on FY2019 authorization and appropriations legislation for NASA.

Figure 1 shows NASA's total annual budget authority from the agency's establishment in FY1958 to FY2018, in both current dollars and inflation-adjusted FY2018 dollars.

---

[1] FY2016 Education amounts are not shown in the FY2018 congressional budget justification and are instead taken from the explanatory statement for the Consolidated Appropriations Act, 2016 (P.L. 114-113), *Congressional Record*, December 17, 2015, pp. H9741-H9743.

Table 1. NASA Appropriations, FY2013-FY2018 (budget authority in $ millions)

	FY2013	FY2014	FY2015	FY2016	FY2017	FY2018
Science	$4,782	$5,148	$5,243	$5,584	$5,765	$6,222
Earth Science	1,659	1,825	1,784	1,927	1,921	1,921
Planetary Science	1,275	1,346	1,447	1,628	1,846	2,228
Astrophysics	617	678	731	762	750	850
James Webb Space Telescope	628	658	645	620	569	534
Heliophysics	603	641	636	647	679	689
Aeronautics	530	566	642	634	660	685
Space Technology	615	576	600	686	687	760
Exploration	3,706	4,113	3,543	3,996	4,324	4,790
Exploration Systems Development	2,884	3,115	3,212	3,641	3,929	4,395
Orion	*1,114*	*1,197*	*1,190*	*1,270*	*1,350*	*1,350*
Space Launch System	*1,415*	*1,600*	*1,679*	*1,972*	*2,150*	*2,150*
Exploration Ground Systems	*355*	*318*	*343*	*399*	*429*	*895*
Commercial Spaceflight	525	696	—[a]	—[a]	—[a]	—[a]
Exploration R&D	297	302	331	355	395	395
Space Operations	3,725	3,774	4,626	5,032	4,951	4,752
Space Shuttle	39	0	8	5	0	0
International Space Station	2,776	2,964	1,525	1,436	n/s	n/s
Space Transportation	—[b]	—[b]	2,254	2,668	n/s[c]	n/s
Space and Flight Support	910	810	839	923	n/s	n/s
Education	116	117	119	115	100	100
Space Grant	37	40	40	40	40	40
EPSCoR	17	18	18	18	18	18
MUREP	28	30	32	32	32	32
Other	34	29	29	25	10	10
Safety, Security, and Mission	2,711	2,793	2,755	2,772	2,769	2,827
Services Construction and EC&R	661[d]	522	446	427	470[e]	562
Inspector General	35	38	37	37	38	39
Total	16,879[d]	17,647	18,010	19,285	19,762[e]	20,736

Sources: FY2013-FY2016 from NASA FY2015-FY2018 congressional budget justifications. FY2017 from P.L. 115-31 and explanatory statement, *Congressional Record*, May 3, 2017, pp. H3374-H3375. FY2018 from P.L. 115-141 and explanatory statement, *Congressional Record*, March 22, 2018, pp. H2094-H2096. See text for details.

Notes: Some totals may not add because of rounding. R&D = Research and Development. EPSCoR = Established Program to Stimulate Competitive Research. MUREP = Minority University Research and Education Program. EC&R = Environmental Compliance and Remediation. n/s = not specified.

[a] Included in Space Transportation under Space Operations.

[b] Commercial Crew funded under Exploration. Remainder of Space Transportation included in International Space Station.

[c] Includes up to $1,185 million for Commercial Crew.

[d] Includes $14 million (after sequestration) of supplemental funding from the Disaster Relief Appropriations Act, 2013 (P.L. 113-2) that is not shown in the NASA FY2015 congressional budget justification.

[e] Includes $109 million in emergency funding (from Sec. 540 of P.L. 115-31) for repairs at NASA facilities damaged by natural disasters.

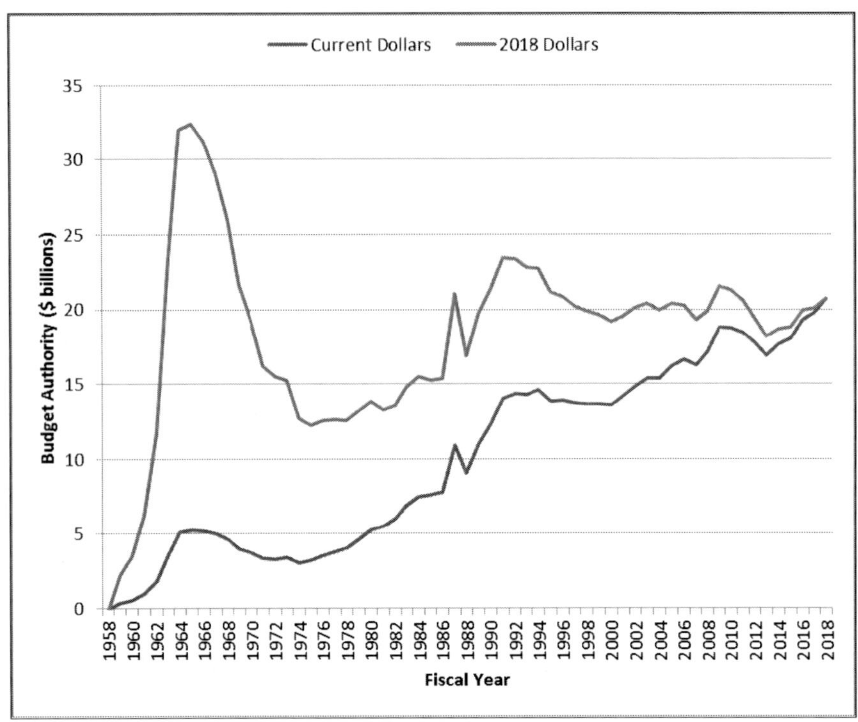

Source: Compiled by CRS. FY1958-FY2008 from National Aeronautics and Space Administration, *Aeronautics and Space Report of the President: Fiscal Year 2008 Activities*, http://history.nasa.gov/presrep2008.pdf, Table D-1A. FY2009-FY2012 from NASA congressional budget justifications, FY2011-FY2014. FY2013-FY2018 as in Table 1. Current dollars deflated to FY2018 dollars using GDP (chained) price index from President's budget for FY2019, Historical Table 10.1, https://www.whitehouse.gov/wp-content/uploads/2018/02/hist10z1-fy2019.xlsx.

Note: Transition quarter between FY1976 and FY1977 not shown.

Figure 1. NASA Funding, FY1958-FY2018.

Table 2. NASA Appropriations, FY2019 (budget authority in $ millions)

	FY2017 Comparable	FY2019 Request	House	Senate	Enacted
Science	$5,762	$5,895			
Earth Science	1,908	1,784			
Planetary Science	1,828	2,235			
Astrophysics[a]	1,352	1,185			
Heliophysics	675	691			
Aeronautics	656	634			
Exploration Research and Technology[b]	827	1,003			
Deep Space Exploration Systems[c]	4,184	4,559			
Exploration Systems Development	3,929	3,670			
- Orion	*1,330*	*1,164*			
- Space Launch System	*2,127*	*2,078*			
- Exploration Ground Systems	*472*	*428*			
Advanced Exploration Systems	98	889			
Exploration R&D	157	—			
LEO and Spaceflight Operations	4,943	4,625			
International Space Station	1,451	1,462			
Space Transportation	2,589	2,109			
Space and Flight Support	903	904			
Commercial LEO Development	—	150			
Education	100	0			
Space Grant	40	0			
EPSCoR	18	0			
MUREP	32	0			
Other	10	0			
Safety, Security, and Mission Services	2,769	2,750			
Construction and EC&R	485d	388			
Inspector General	38	39			
Total	19,762[d]	19,892			

Source: FY2019 NASA congressional budget justification and P.L. 115-31. See text for details.

Notes: Some totals may not add because of rounding. For account structure changes, see table notes and discussion in text. R&D = Research and Development. LEO = Low Earth Orbit. EPSCoR = Established Program to Stimulate Competitive Research. MUREP = Minority University Research and Education Program. EC&R = Environmental Compliance and Remediation. n/s = not specified.

[a] Includes the James Webb Space Telescope, formerly a separate item.

[b] Formerly Space Technology, plus elements formerly in Exploration.

[c] Formerly Exploration, minus elements now in Exploration Research and Technology.

[d] Includes $109 million in emergency funding (from Sec. 540 of P.L. 115-31, not shown in the FY2019 NASA congressional budget justification) for repairs at NASA facilities damaged by natural disasters.

In: Science Policies and Programs
Editor: Johnnie Rodgers
ISBN: 978-1-53614-107-8

Chapter 2

DEFENSE SCIENCE AND TECHNOLOGY FUNDING*

John F. Sargent Jr.

ABSTRACT

Defense science and technology (Defense S&T) is a term that describes a subset of Department of Defense (DOD) research, development, testing, and evaluation (RDT&E) activities. The Defense S&T budget is the aggregate of funding provided for the three earliest stages of DOD RDT&E: basic research, applied research, and advanced technology development. Defense S&T is of particular interest to Congress due to its perceived value in supporting technological advantage and its importance to key private sector and academic stakeholders.

Advocates of strong and sustained Defense S&T funding assert that Defense S&T funding plays important and unique roles in the DOD innovation system, supporting medium-term, evolutionary technologies and incremental innovation that help improve existing products and systems, as well as longer-term, revolutionary technologies providing

---

* This is an edited, reformatted, and augmented version of a Congressional Research Service report, R45110, dated February 21, 2018.

U.S. technological dominance, deterring conflict, and, when necessary, defeating adversaries. Both evolutionary and revolutionary technologies are viewed by most warfighters and policymakers as central to U.S. national security as well as to the lives of those serving in uniform.

In FY2017, Defense S&T was $13.4 billion, nearly six times the FY1978 level of $2.3 billion. Most growth occurred from FY1978 to FY2006, at a compound annual growth rate (CAGR) of 6.4%. From FY2006 to FY2017, growth was slower (0.1% CAGR). Most of the growth and volatility was in advanced technology development. In FY2017 constant dollars, Defense S&T funding peaked at $16.2 billion in FY2005 and declined by $2.8 billion through FY2017.

In FY2016, basic research accounted for $2.2 billion of the Defense S&T total. The Navy accounted for the largest share of DOD basic research (29.2%), followed by the Defense-Wide agencies (27.6%), Air Force (23.0%), and Army (20.3%). Universities and colleges performed nearly half ($1.1 billion, 48.8%) of DOD basic research in FY2016; DOD and other intramural federal laboratories performed 22.9%; industry, 18.2%; other non-profits, 7.5%; federally funded research and development centers (FFRDCs), 0.7%; and others, 2.0%.

A number of recommendations have been put forth by various organizations regarding the appropriate level of funding for Defense S&T and DOD basic research, as well as the level of funding for investments in research supporting potentially revolutionary advancements.

A 1998 Defense Science Board (DSB) report recommended setting Defense S&T at 3.4% of total DOD funding. In 2001, the Quadrennial Defense Review recommended that 3.0% of total DOD funding be directed toward Defense S&T. In FY1996, Defense S&T was at the 3.0% level. It subsequently fell to 1.7% in FY2011 and has since risen to 2.2%. An alternative approach recommended by the DSB in 1998 was to set Defense S&T at a percentage of DOD RDT&E, similar to the industry ratio of research funding to total R&D funding (which it calculated for the pharmaceutical industry as 24%). In 2015, the Coalition for National Security Research (CNSR), a coalition of industry, universities, and associations, recommended a target of 20%. At the time of the DSB report, S&T's share of DOD RDT&E was approximately 21%. After rising to 21.5% in FY2000, Defense S&T's share fell to 15.2% in FY2011, and then rose to 17.9% in FY2016.

With respect to DOD basic research, the Council on Competitiveness (2004) and the CNSR (2015) recommended a target of at least 20% of Defense S&T. As a share of Defense S&T, basic research declined from 14.6% in FY1996 to 11.0% in FY2006, then began a steady rise to 18.4% in FY2015. In FY2016, basic research's share of Defense S&T was 17.4%. In its 1998 report, the DSB recommended that one-third of Defense S&T be devoted to research targeted toward revolutionary technological advancements. The Defense Advanced Research Projects

Agency (DARPA) has been the lead DOD agency focused on revolutionary R&D. In FY2017, DARPA accounted for 21.6% of Defense S&T.

This chapter provides an overview of the portion of Department of Defense (DOD) research, development, testing, and evaluation (RDT&E) funding referred to as Defense Science and Technology (Defense S&T). It provides perspectives on the role of Defense S&T in supporting U.S. defense capabilities, historical funding levels, recent funding trends, and approaches to determining how much the federal government should invest in Defense S&T, particularly in basic research.[1]

WHAT IS DEFENSE SCIENCE AND TECHNOLOGY?

Congress provides appropriations to DOD for RDT&E activities conducted in support of its mission requirements. DOD's *Financial Management Regulation* (DOD 7000.14-R) provides a taxonomy for requesting, tracking, and accounting for federal investments in RDT&E based on the character of work performed. DOD budget justifications and congressional appropriations reports and explanatory statements typically employ this taxonomy which consists of seven budget activity codes (6.1 through 6.7) and a description (as shown in Table 1).[2]

---

[1] Issues related to overall DOD RDT&E are addressed in CRS Report R44711, *Department of Defense Research, Development, Test, and Evaluation (RDT&E): Appropriations Structure*, by John F. Sargent Jr., CRS In Focus IF10553, *Defense Primer: RDT&E*, by John F. Sargent Jr., and CRS Report R45088, *Defense Advanced Research Projects Agency: Overview and Issues for Congress*, by Marcy E. Gallo. This report addresses issues specifically related to Defense S&T and basic research.

[2] DOD RDT&E appropriations acts typically do not include this taxonomy, instead making appropriations to the Defense-Wide, Army. Navy, Air Force, and other accounts; these appropriations support specific program elements associated with the budget activity codes. For example, funding for Army RDT&E is provided in Title IV (Research, Development, Test, and Evaluation), Research, Development, Test, and Evaluation, Army. Appropriations Committee, and conference reports, as well as explanatory statements, generally provide more detailed guidance for the use of these funds. For example, within the Army RDT&E appropriation of $8.3 billion, the 2017 explanatory statement identifies specific program elements and Congress' intended funding levels (e.g., In-House Laboratory Independent Research (program element 0601101A, a part of Army basic research, budget activity code 1), $12.4 million).

Table 1. DOD RDT&E Budget Activity Codes and Descriptions

Budget Activity Code	Description
6.1	Basic Research
6.2	Applied Research
6.3	Advanced Technology Development
6.4	Advanced Component Development and Prototypes
6.5	System Development and Demonstration
6.6	RDT&E Management Support
6.7	Operational Systems Development

Source: Department of Defense, *Financial Management Regulation (DOD 7000.14-R),* Volume 2B, March 2016, http://comptroller.defense.gov/fmr.aspx.

Defense Science and Technology is a subset of DOD RDT&E appropriations that includes funding for basic research (6.1), applied research (6.2), and advanced technology development (6.3)—the earliest stages of the RDT&E process.

DOD defines these budget activities in the following manner:

> Basic research [Budget Activity Code 6.1] is systematic study directed toward greater knowledge or understanding of the fundamental aspects of phenomena and of observable facts without specific applications towards processes or products in mind. It includes all scientific study and experimentation directed toward increasing fundamental knowledge and understanding in those fields of the physical, engineering, environmental, and life sciences related to long-term national security needs. It is farsighted high payoff research that provides the basis for technological progress. Basic research may lead to: (a) subsequent applied research and advanced technology developments in Defense-related technologies, and (b) new and improved military functional capabilities in areas such as communications, detection, tracking, surveillance, propulsion, mobility, guidance and control, navigation, energy conversion, materials and structures, and personnel support.
>
> Applied research [Budget Activity Code 6.2] is systematic study to understand the means to meet a recognized and specific need. It is a systematic expansion and application of knowledge to develop useful materials, devices, and systems or methods. It may be oriented,

ultimately, toward the design, development, and improvement of prototypes and new processes to meet general mission area requirements. Applied research may translate promising basic research into solutions for broadly defined military needs, short of system development. This type of effort may vary from systematic mission-directed research beyond that in Budget Activity 1 [basic research] to sophisticated breadboard hardware, study, programming and planning efforts that establish the initial feasibility and practicality of proposed solutions to technological challenges. It includes studies, investigations, and non-system specific technology efforts. The dominant characteristic is that applied research is directed toward general military needs with a view toward developing and evaluating the feasibility and practicality of proposed solutions and determining their parameters. Applied research precedes system specific technology investigations or development.

Advanced Technology Development, [Budget Activity Code 6.3] includes development of subsystems and components and efforts to integrate subsystems and components into system prototypes for field experiments and/or tests in a simulated environment. Budget Activity 3 includes concept and technology demonstrations of components and subsystems or system models. The models may be form, fit, and function prototypes or scaled models that serve the same demonstration purpose. The results of this type of effort are proof of technological feasibility and assessment of subsystem and component operability and producibility rather than the development of hardware for service use. Projects in this category have a direct relevance to identified military needs. Advanced Technology Development demonstrates the general military utility or cost reduction potential of technology when applied to different types of military equipment or techniques....Projects in this category do not necessarily lead to subsequent development or procurement phases, but should have the goal of moving out of Science and Technology (S&T) and into the acquisition process within the Future Years Defense Program (FYDP). Upon successful completion of projects that have military utility, the technology should be available for transition.[3]

[3] Department of Defense, *Financial Management Regulation (DoD 7000.14-R)*, Volume 2B, March 2016.

PERSPECTIVES ON THE ROLES AND VALUE OF DEFENSE S&T

Defense S&T is of particular interest and importance to Congress due to its perceived value in supporting military competitive advantage. Defense S&T is also of interest to key stakeholders in the private sector and academia. For example, advocates of strong and sustained Defense S&T funding assert that this funding plays important and unique roles in the DOD innovation system. The scientific and technological insights that emerge from Defense S&T funding—often referred to as the nation's "seed corn"—are seen by many as the critical body of knowledge available to DOD and the industrial base for future defense technology development.[4] Defense S&T supports both:

- medium-term, evolutionary technologies and incremental innovations to help improve existing products and systems; and
- longer-term, revolutionary technologies to support U.S. technological dominance, deter conflict, and defeat adversaries.

These technologies—both evolutionary and revolutionary—are seen by most warfighters and policymakers as central to U.S. national security as well as to the lives of those serving in uniform in the medium and long term.

In contrast, most of the balance of DOD RDT&E is focused on near-term applications. Budget activity 6.4, Advanced Component Development and Prototypes, efforts are directed toward the evaluation of integrated technologies and prototype systems in realistic operating environments, not just in controlled laboratory environments. Funding in this budget activity seeks to expedite technology transition from the laboratory to operational use. Budget activity 6.5, System Development and Demonstration, is

---

[4] Seed corn refers to the high quality kernels of corn (and other crops) used as seeds for growing future crops. Thus, seed corn has been essential to maintaining agricultural output. The term is used metaphorically to refer to an asset or investment that is expected to provide future returns.

engineering and manufacturing development tasks aimed at meeting validated requirements prior to full-rate production. At this stage, prototype performance is near or at planned operational system levels. Budget activity 6.7, Operational Systems Development, is focused on development efforts to upgrade systems that have been fielded or have received approval for full rate production and anticipate production funding in the current or subsequent fiscal year. Budget activity 6.6, RDT&E Management Support, includes management support for RDT&E efforts and funds to sustain and/or modernize the installations required for general research. Accordingly, BA 6.6 funding supports RDT&E activities in each of the other budget activities.[5]

From FY2007 to FY2017, Defense S&T averaged 17.1%, approximately one-sixth, of total Defense RDT&E (ranging from 15% to 19% during these years). Historical funding levels and recent trends are discussed in more detail in the following section of this chapter.

According to the National Academies' 2007 report *Rising Above the Gathering Storm*:

> Keeping a technological edge over adversaries of the United States has long been a key component of our national security strategy. US preeminence in science and technology is considered essential to achieving that goal.[6]

The report further emphasizes the importance of DOD basic research, asserting that

> The importance of DOD basic research is illustrated by its products—in defense areas these include night vision; stealth technology; near-real-time delivery of battlefield information; navigation, communication, and weather satellites; and precision munitions.[7]

---

[5] Department of Defense, *Financial Management Regulation (DoD 7000.14-R)*, Volume 2B, March 2016.

[6] National Academies, *Rising Above the Gathering Storm: Energizing and Employing America for a Brighter Economic Future*, 2007, p. 483.

[7] Ibid, p. 140. The report also asserts the importance of Defense basic research for its contribution to non-defense applications, stating "The Internet, communications and weather satellites,

DOD investments in basic research are also considered vital to maintaining university research, the education of scientists and engineers, and the preservation of teaching capacity in key scientific and engineering fields. Proponents of these investments assert that it is essential to ensure steady funding to these fields to ensure stability for professors, researchers, and academic programs. Uneven funding patterns, some assert, can create uncertainty (in positions, salary, equipment, and programs, for example) that may drive out some of the best scientists and engineers and discourage the most capable students from pursuing degrees and research in these disciplines, resulting in adverse impacts on future innovation in fields key to national security.

Some analysts express concern that, in times of tightly constrained budgets, Defense S&T may be an easy target for budget cuts. Cuts to Defense S&T might produce few short-term consequences to national defense, as the benefits of these investments tend to be realized in the medium- to long-term. However, the neglect of these earlier-stage research and development activities could have serious medium- and long-term consequences, depriving the U.S. defense sector of the critical underpinnings necessary for maintaining technological superiority and global dominance in the future.[8]

Former Under Secretary of Defense for Acquisition, Technology and Logistics Frank Kendall noted:

> R&D is not a variable cost. R&D drives our rate of modernization. It has nothing to do with the size of the force structure. So, when you cut R&D, you are cutting your ability to modernize on a certain time scale, period—no matter how big your force structure is ... [T]he investments we're making now in technology are going to give us the forces that we're going to have in the future. The forces we have now came out of

---

global positioning technology, the standards that became JPEG, and even the search technologies used by Google all had origins in DOD basic research" and asserts that these investments are "the primary reason that the United States leads the world today in information technology."

[8] See, for example, Aviation Week and Space Technology, "Budget Cuts to Future Weapons Could Have Long-Term Impact," April 3, 2015, http://aviationweek.com/technology/budget-cuts-future-weapons-could-have-long-term-impact.

> investments that were made, to some extent, in the 80s and 90s...if you give up the time it takes for lead time to get...a capability, you are not going to get that back.[9]

Alan R. Shaffer, Principal Deputy, Assistant Secretary of Defense for Defense Research and Engineering, underscored this point, stating:

> If we don't do the research and development for a new system then the number of systems of that type we will have is zero. It is not variable.[10]

Such cuts may also result in lasting damage to important segments of the U.S. R&D infrastructure—researchers, professors, academic programs, student interest, equipment, infrastructure, etc.—in defense-critical fields, even if funding were to be later restored. Such effects could not only diminish U.S. innovative capacity, but result in the transfer of knowledge and loss of people, capabilities, and leadership to other nations. According to a 2012 Defense Science Board report on DOD basic research:

> The DOD basic research program has supported a large fraction of revolutionary research in the physical sciences. Without DOD support, these U.S.-based research communities would find it more difficult to expand knowledge, collaborate, publish, and meet. Without adequate U.S. support, these centers of knowledge will drift to other countries.[11]

While there is little direct opposition to Defense S&T spending in its own right, there is intense competition for available dollars in the appropriations process. This competition has been made more acute under

---

[9] Honorable Frank Kendall, Under Secretary of Defense for Acquisition, Technology and Logistics, presentation to McAleese/Credit Suisse FY 2015 Defense Programs Conference, February 25, 2014, as cited in testimony of Alan R. Shaffer, Principal Deputy, Assistant Secretary of Defense for Defense Research and Engineering, before the Senate Armed Services Committee, Subcommittee on Emerging Threats and Capabilities, April 8, 2014.

[10] Testimony of Alan R. Shaffer, Principal Deputy, Assistant Secretary of Defense for Defense Research and Engineering, before the Senate Armed Services Committee, Subcommittee on Emerging Threats and Capabilities, April 8, 2014.

[11] Department of Defense, Defense Science Board, *Report of the Defense Science Board Report on Basic Research,* January 2012, p. 9, https://www.acq.osd.mil/dsb/reports/2010s/BasicResearch.pdf.

congressionally enacted budget control provisions.[12] Congressional acts establish and provide enforcement mechanisms for separate spending caps for defense and non-defense spending. These independent budget caps essentially fence off certain funds from being used for defense purposes by those who would prioritize such defense spending over certain non-defense activities. Increases in the defense and non-defense budget caps for FY2018 and FY2019 included in the Bipartisan Budget Act of 2018 (P.L. 115-123) may ease the resource competition in these fiscal years, but not eliminate it. With the spending caps has come greater competition for available dollars within the defense portion of the budget (for example, between RDT&E and procurement), and among the various RDT&E budget activities and program elements (for example, between Defense S&T and the rest of the defense RDT&E budget activities). With members of the U.S. Armed Forces currently engaged in combat and others facing potentially imminent threats in other locations around the world, some may believe it is appropriate to prioritize defense spending to support immediate operational needs and contingency preparations of the military over activities whose payoff is likely to be realized only in the longer term.

In addition, some have questioned the effectiveness of defense investments in R&D. For example, a 2012 article published by the Center for American Progress (CAP), a public policy research and advocacy organization, notes that the technological superiority of the United States did not initially provide an effective defense for U.S. troops against low-tech improvised explosive devices (IEDs) in Iraq and Afghanistan. The article also asserts that many high-priced major weapons systems—such as President Reagan's missile defense program—have failed to deliver on their promised capabilities due to scientific and engineering shortcomings. Further the article notes that commercial technology development is now outstripping defense technology due to the "strength of capitalism"—including large markets, consumer demand, and competitive challenges—

---

[12] See, for example, the Budget Control Act of 2011 (BCA, P.L. 112-25), American Taxpayer Relief Act of 2012 (P.L. 112-240), and the Bipartisan Budget Act of 2013 (P.L. 113-67), Bipartisan Budget Act of 2015 (P.L. 114-74). For more information on the BCA and the federal budget, see CRS Report R42506, *The Budget Control Act of 2011 as Amended: Budgetary Effects*, by Grant A. Driessen and Marc Labonte.

suggesting the potential benefits of pursuing a technology acquisition strategy based more heavily on off-the-shelf technologies or the repurposing of those technologies to meet defense needs. The article treats basic research less harshly than other Defense RDT&E activities, which CAP describes as "the kind of boondoggle R&D spending the Pentagon engages in at the applied and developmental level."[13]

HISTORICAL DEFENSE S&T FUNDING AND RECENT TRENDS

Defense S&T has grown substantially in current dollars (unadjusted for changes in buying power) over the past four decades, from $2.3 billion in FY1978 to $13.4 billion in FY2017.[14] This growth is illustrated in Figure 1, which shows this growth by its component budget activities. During the FY1978-FY2017 period, Defense S&T grew at a compound annual growth rate (CAGR) of 4.6%. Most of this growth occurred between FY1978 and FY2006 (6.4% CAGR); from FY2006 to FY2017 Defense S&T grew at a pace of 0.1% CAGR, though the period was punctuated by periods of growth and contraction. The funding trends for each of the component budget activities (6.1-6.3) during the FY1978-FY2017 period were different.

- Basic research (6.1) funding grew at 4.4% CAGR from FY1978 to FY2017, approximately the same pace as overall Defense S&T funding (4.7% CAGR), but the growth was steadier, with fewer periods of substantial decrease.
- Applied research (6.2) funding grew steadily from FY1978 to FY2017, in general, but at a slightly slower rate (3.5% CAGR)

[13] Eric Altman, Senior Fellow, *Is Defense R&D Spending Effective?*, Center for American Progress, January 13, 2012, https://www.americanprogress.org/issues/general/news/2012/01/13/11001/think-again-is-defense-rd-spendingeffective.

[14] FY2018 and FY2019 request levels are included in several of the charts in this section for reader reference, but are not discussed in the analyses.

than overall Defense S&T and basic research funding. Similar to overall Defense S&T, most of the growth in applied research occurred between FY1978 and FY2006 (4.8% CAGR); from FY2006 to FY2017 applied research grew at a slower pace of 0.3% CAGR.

- Advanced technology development (6.3) funding experienced periods of growth and decline from FY1978 to FY2017. From FY1978 to FY1993, advanced technology development grew at a rate of 14.4% CAGR. From FY1993 to FY1999, funding declined at a rate of 3.0% CAGR. Funding grew at a rate of 10.3% CAGR from FY1999 to FY2006, and then fell again from FY2006 to FY2013 at a rate of 4.9% CAGR. Most recently, funding for advanced technology development has grown at a rate of 5.9% CAGR from FY2013 to FY2017.

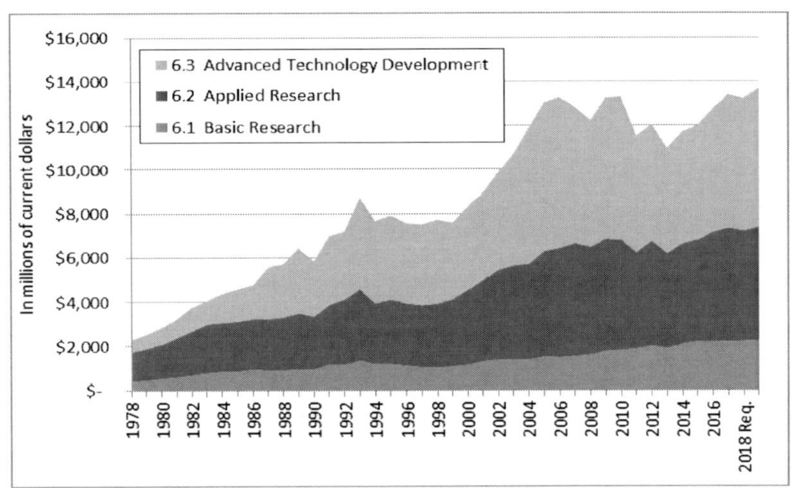

Source: CRS analysis of data from Department of Defense, *Research, Development, Test, and Evaluation Programs (R-1)* for FY1978-2019. CRS used funding levels from two years before the request year. For example, the FY2017 funding levels are from the FY2019 R-1. FY2018 and FY2019 data are request levels from the FY2018 R-1 and FY2019 R-1, respectively.

Notes: FY1978-FY2017 (actual), FY2018 (request), FY2019 (request). Req.=Request.

Figure 1. Defense S&T Funding, by Budget Activity, FY1978-FY2019; In millions of current dollars.

Figure 2 illustrates Defense S&T by budget activity in constant FY2017 dollars.[15] This figure provides an illustration of Defense S&T funding levels from FY1978 to FY2017 in terms of the purchasing power of these funds.[16]

Defense S&T grew by nearly 90% in constant dollars between FY1978 and FY2017. Despite the increase, there were periods of decline. Between FY1993 and FY1999, funding decreased at a rate of 4.0% CAGR. Funding rebounded between FY1999 and FY2005 (when Defense S&T funding reached its peak in constant dollars for the FY1978-FY2017 period), growing by 7.1% CAGR. This growth period was followed by another period of decline through FY2013 (4.0% CAGR). From FY2013 to FY2017, Defense S&T grew at a rate of 3.5% CAGR.

- Basic research funding grew, with some ups and downs, at a rate of 1.4% CAGR, during the FY1978-FY2017 period. From FY1993 to FY1998, funding fell by nearly 30%, then recovered, surpassing its FY1993 level in FY2012. Funding then rose an additional 1.4% between FY2012 and FY2017.
- Applied research funding was essentially flat through FY1998, then grew steadily through FY2005 (5.1% CAGR) and remained flat again through FY2007. Funding fell by 24% from FY2007 to FY2013, declining at a rate of 4.5% CAGR. Funding recovered between FY2013 and FY2017, rising by 13.4%, at a rate of 3.2% CAGR.
- The largest swings in Defense S&T resulted from changes in the advanced technology development funding component. Advanced technology development funding nearly quadrupled in constant dollars from FY1978 to FY1993. From FY1993 through FY1999,

---

[15] As calculated using the GDP (Chained) Price Index from Table 10.1 of the Historical Tables in the President's Budget for Fiscal Year 2019, to adjust for inflation; this index is used by the Office of Management and Budget to convert federal research and development outlays from current dollars to constant dollars. https://www.whitehouse.gov/sites/whitehouse.gov/files/omb/budget/fy2018/hist10z1.xls.

[16] The President's FY2018 and FY2019 request levels are also included in this figure for reference.

it fell by 25%, before rising again to its constant dollar peak in FY2005. Between FY2005 and FY2013, advanced technology development fell by 39%, and then recovered somewhat between FY2013 and FY2017 (up 19%).

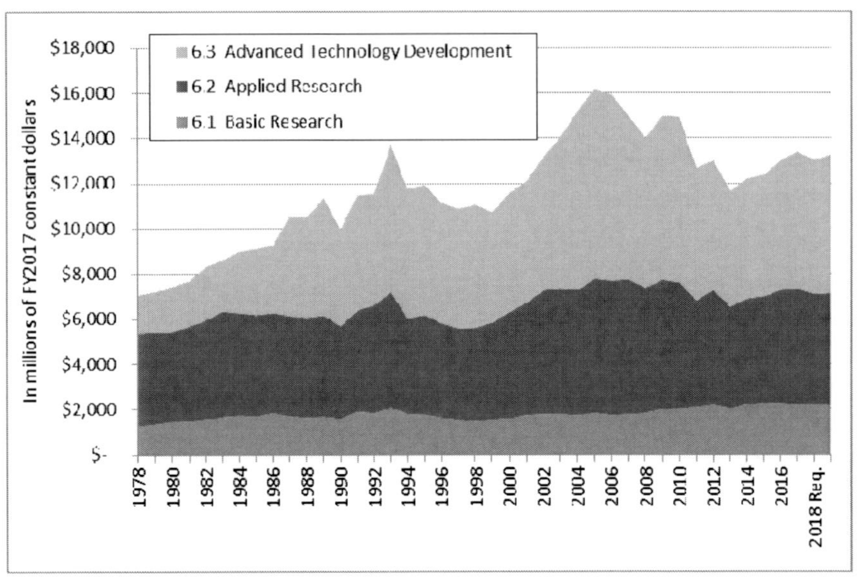

Source: CRS analysis of data from Department of Defense, *Research, Development, Test, and Evaluation Programs (R-1)* for FY1978-2019. CRS used funding levels from two years before the request year. For example, the FY2017 funding levels are from the FY2019 R-1. FY2018 and FY2019 data are request levels from the FY2018 R-1 and FY2019 R-1, respectively.

Notes: FY1978-FY2017 (actual), FY2018 (request), FY2019 (request). For purposes of this chart, CRS used the GDP (Chained) Price Index from Table 10.1 of the Historical Tables in the President's Budget for Fiscal Year 2019, to adjust for inflation; this index is used by the Office of Management and Budget to convert federal research and development outlays from current dollars to constant dollars. https://www.whitehouse.gov/wp-content/uploads/2018/02/hist10z1-fy2019.xlsx.

Req.=Request.

Figure 2. Defense S&T Funding, by Budget Activity, FY1978-FY2019; In millions of constant FY2017 dollars.

DOD BASIC RESEARCH FUNDING

In FY2016, DOD spent an estimated $2.3 billion on basic research. The following sections describe the composition of DOD basic research funding by organizational component and the composition of performers of the research by organizational component.[17]

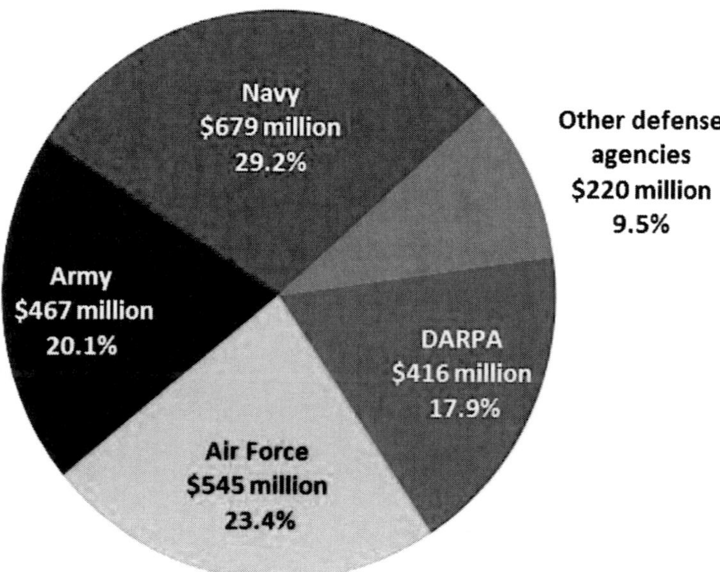

Source: CRS analysis of data from National Science Foundation, National Center for Science and Engineering Statistics, *Survey of Federal Funds for Research and Development, FYs 2015–17*, Table 31.

Notes: According to NSF, FY2016 data are estimates of congressional appropriations actions and apportionment and reprogramming decisions. Percentages rounded and therefore may not add to100%.

Figure 3. Basic Research Funding of DOD Services and Agencies, FY2016.

[17] FY2016 is the latest available data for analysis of DOD basic research by performing sector.

DOD Basic Research Funding by Organizational Component

The Department of Defense funds basic research activities through the Army, Navy, Air Force, and Defense-Wide (D-W) agencies.[18] Figure 3 illustrates the composition of that funding based on FY2016 obligations. Funding was broadly distributed with each of the services and Defense-Wide agencies accounting for 20%-30% of total DOD basic research funding. The Navy accounted for the largest share (29.2%) of DOD basic research, followed by the Defense-Wide agencies (27.3%), the Air Force (23.4%), and the Army (20.1%). The Defense Advanced Research Projects Agency (DARPA) is the largest funder of basic research among the D-W agencies, accounting for 17.9% of total DOD RDT&E.

Basic Research of DOD Components by Performing Sector

Figure 4 illustrates the share of total DOD basic research by performing sector.[19] Universities and colleges performed nearly half ($1.1 billion, 48.8%) of DOD basic research in FY2016. Nearly another quarter of DOD basic research was performed by intramural performers ($533 million, 22.9%). Industry performed 18.2% ($423 million) of DOD basic

---

[18] Defense-Wide agencies engaged in DOD RDT&E include, but are not limited to, the Defense Advanced Research Projects Agency (DARPA), Office of the Secretary of Defense (OSD), Missile Defense Agency (MDA), Chemical and Biological Defense Program (CBDP), Defense Contract Management Agency (DCMA), Defense Human Resources Activity (DHRA), Defense Information Systems Agency (DISA), Defense Logistics Agency (DLA), Defense Security Cooperation Agency (DSCA), Defense Security Service (DSS), Defense Technical Information Center (DTIC), Defense Threat Reduction Agency (DTRA), Operational Test and Evaluation, the Joint Staff (TJS), U.S. Special Operations Command (SOCOM), and Washington Headquarters Service (WHS).

[19] Data collected by the National Science Foundation through the *Survey of Federal Funds for Research and Development* assigns obligations to nine performing sectors: intramural; universities and colleges; industry; federally funded research and development centers (by type of administrator: industry, university, and nonprofit); state and local governments; other nonprofits; and foreign. Intramural research is that performed by federal employees using federally owned and operated facilities. Research performed by all other performers is characterized as extramural research. For more information see the technical notes to the *Survey of Federal Funds for Research and Development Fiscal Years 2015–17*, which can be accessed at https://ncsesdata.nsf.gov/fedfunds/2015/fedfunds_2015_tech_notes.pdf.

research; other nonprofits 7.5% ($174 million); and all other performers 1.7% ($61 million).

Figure 5 illustrates the composition of DOD components' basic research by performing sector.[20] As the charts show, the components' degree of reliance on performing sectors varies. In FY2016:

- All components relied heavily on universities and colleges, especially the other defense agencies (59.2%).
- Reliance on industry varied widely, from 40.6% (DARPA) to 5.0% (Navy).

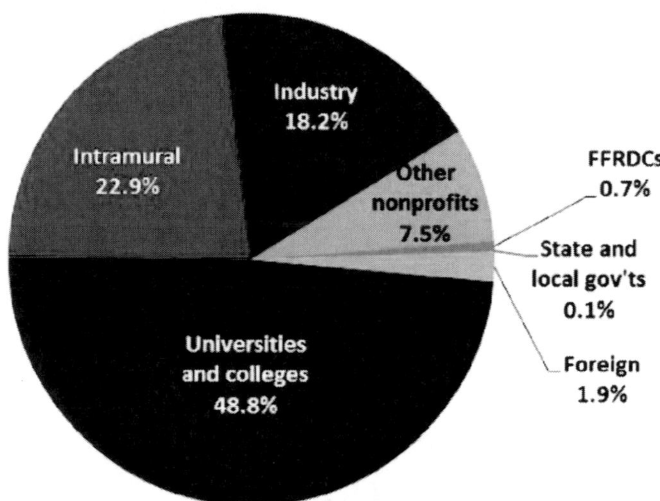

Source: CRS analysis of data from National Science Foundation, National Center for Science and Engineering Statistics, *Survey of Federal Funds for Research and Development, FYs 2015–17*, Table 31.

Notes: According to NSF, FY2016 data are estimates of congressional appropriations actions and apportionment and reprogramming decisions. FFRDC=federally funded research and development center.

Figure 4. DOD Basic Research Obligations by Performing Sector, FY2016.

[20] DARPA and the other defense agencies comprise the D-W agencies.

- The components' reliance on intramural performers also varied, from 5.2% (DARPA) to 32.7% (Navy).
- Other nonprofits are significant performers for the Air Force (14.3%) and DARPA (13.2%), but barely used by the Army (0.1%) and other defense agencies (1.3%).

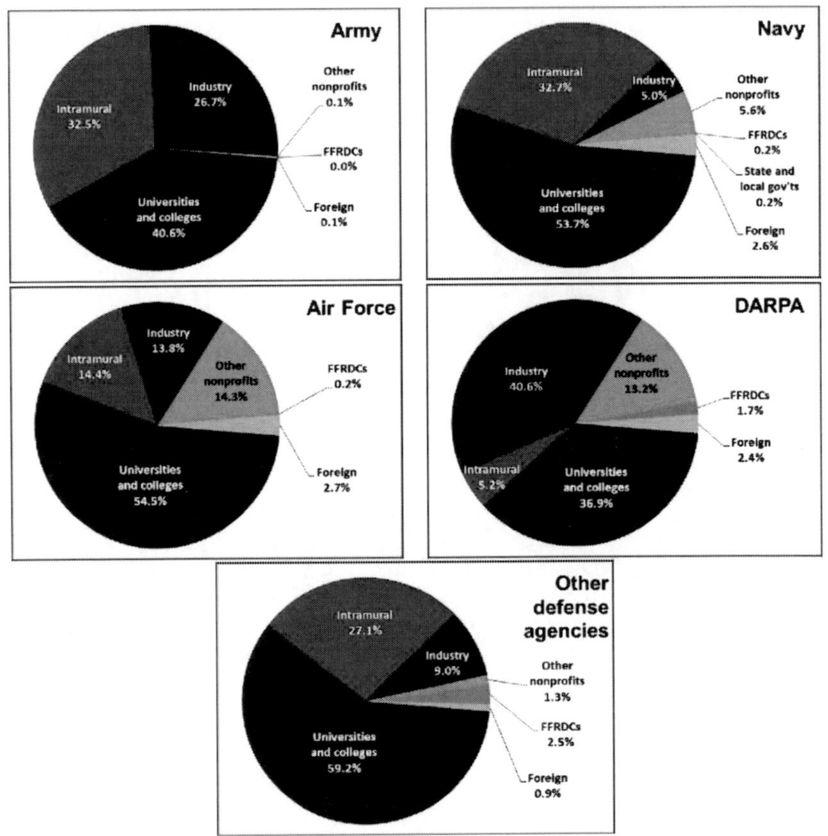

Source: CRS analysis of data from National Science Foundation, National Center for Science and Engineering Statistics, *Survey of Federal Funds for Research and Development, FYs 2015–17*, Table 31.

Notes: According to NSF, FY2016 data are estimates of congressional appropriations actions and apportionment and reprogramming decisions. FFRDC=Federally funded research and development center.

Figure 5. Share of DOD Components' Basic Research Obligations by Performing Sector, FY2016.

Program Elements in DOD Basic Research

According to DOD,

> The program element is the primary data element in the Future Years Defense Program (FYDP) and normally the smallest aggregation of resources used by the Office of the Secretary of Defense [OSD] for analysis. It generally represents a collection of functional or organizational entities and their related resources. PEs are designed and quantified to be comprehensive and mutually exclusive. As the building blocks of the programming and budgeting system, PEs are continually reviewed to maintain proper visibility into the multitude of defense programs.[21]
>
> DOD RDT&E is generally requested and funded under specified program elements (PEs). Each program element is associated with a seven character number and an alphanumeric suffix which, in part, indicate the budget activity code and the DOD department or agency receiving the funds. Table 2 identifies each of the basic research PEs for the services and Defense-Wide agencies, their FY2017 enacted funding levels, and their share of each component's basic research funding.

DOD basic research has some program elements that are continuing efforts (that is, they continue across multiple fiscal years). In addition, some of the PEs are common to one or more of the services or Defense-Wide agencies.

The following section provides descriptions of these PEs, as well as the other basic research PEs. The descriptions are drawn largely from the FY2018 budget justifications of the services and Defense-Wide agencies.

Defense Research Sciences

The Defense Research Sciences programs conducted by the Army, Navy, Air Force, and DARPA comprise the largest component and the core of the DOD basic research program.

---

[21] Defense Acquisition University, Department of Defense, website, "Future Years Defense Program (FYDP)," https://www.dau.mil/acquipedia/Pages/ArticleDetails.aspx?aid=a2cc2ade-6336-433e-a088-42f497cdf7ef.

Table 2. Basic Research Program Elements, Services and D-W Agencies, FY2017; Dollars in millions

Program Element	Army		Navy		Air Force		Defense-Wide		Total
	Dollars	Share of Army 6.1	Dollars	Share of Navy 6.1	Dollars	Share of Air Force 6.1	Dollars	Share of D-W 6.1	
Defense Research Sciences	$286.1	60.5%	$413.8	75.3%	$370.6	71.1%	$432.3	62.0%	1,502.9
University Research Initiatives	66.5	14.1%	117.3	21.4%	137.8	26.4%			321.6
In-House Laboratory Indep. Research	11.9	2.5%	18.2	3.3%					30.2
Univ. and Industry Research Centers	108.7	23.0%							108.7
High Energy Laser Research Initiatives					13.2	2.5%			13.2
National Defense Education Program							74.3	10.7%	74.3
Basic Research Initiatives							40.6	5.8%	40.6
Basic Operational Medical Research Science							43.1	6.2%	43.1
Chemical and Biological Defense Program							43.9	6.3%	43.9
Defense Threat Reduction Agency							37.2	5.3%	37.2
Univ. Strategic Partnership Basic Res.									
Historically Black Colleges and Universities							25.9	3.7%	25.9
Total	$473.2	100.0%	$549.4	100.0%	$521.6	100.0%	$697.3	100.0%	2,241.5

Source: Department of Defense, Research, Development, Test, and Evaluation Programs (R-!), FY20!9, February 2018.

Note: Column elements may not add to totals due to rounding.

Army

The Army Defense Research Sciences PE supports the development of fundamental scientific knowledge intended to contribute to the sustainment of Army scientific and technological superiority in land warfighting capability and to solving military problems related to long-term national security needs, supports investigation of new concepts and technologies for the Army's future force, and seeks to provide the means to exploit scientific breakthroughs and avoid technological surprises. This PE fosters innovation in Army niche areas (e.g., lightweight armor, energetic materials, and night vision capability) and areas where there is no commercial investment due to limited markets (e.g., vaccines for tropical diseases). It also focuses university single investigator research on areas of high interest to the Army (e.g., high-density compact power and novel sensors). The in-house portion of the program relies on the Army's scientific talent and specialized facilities to transition knowledge and technology into appropriate developmental activities. The extramural program leverages the research efforts of other government agencies, academia, and industry.[22]

Navy

The Navy Defense Research Sciences PE supports development of new technological concepts for the maintenance of naval power and national security, and to prevent scientific surprise. The program seeks to exploit scientific breakthroughs and to provide options for new future naval capabilities and innovative naval prototypes. The basic research efforts include scientific study and experimentation directed toward increasing knowledge and understanding in national security-related aspects of physical, engineering, environmental, and life sciences. The program's investments include National Naval Responsibilities (NNRs)[23] and the Basic Research Challenge Program.[24]

---

[22] Department of Defense, *Department of Defense Fiscal Year (FY) 2018 Budget Estimates, Army Justification Book of Research, Development, Test & Evaluation, Army RDT&E, Volume I, Budget Activity 1*, May 2017.

[23] NNRs are areas that are uniquely important to the Navy and Marine Corps that are not addressed by research investments from other DOD services, other federal R&D funding

Air Force

The Air Force Defense Research Sciences PE funds extramural research activities in academia and industry along with in-house investigations performed in the Air Force Research Laboratory (AFRL). Funding supports fundamental broad-based scientific and engineering research in areas critical to Air Force weapon, sensor, and support systems.[25]

DARPA

The DARPA Defense Research Sciences PE seeks to provide the technical foundation for long-term national security enhancement through the discovery of new phenomena and the exploration of the potential of such phenomena for defense applications. It supports scientific study and experimentation that serves as the basis for more advanced knowledge and understanding in information, electronic, mathematical, computer, biological, and materials sciences.

University Research Initiatives

Army

The Army's University Research Initiatives (URI) PE supports several activities, including the Multidisciplinary University Research Initiative (MURI), the Defense University Research Instrumentation Program (DURIP), the Presidential Early Career Awards for Scientists and Engineers (PECASE) program, and the Army's contribution to the Minerva Research Initiative (MRI). The MURI program supports

---

agencies (such as the National Science Foundation, the National Institutes of Health), or private industry.

[24] The Basic Research Challenge Program competitively funds promising research programs in areas not addressed by the current basic research program, placing a focus on high-risk basic research projects in multidisciplinary and collaborative departmental efforts. Department of Defense, *Department of Defense Fiscal Year (FY) 2018 Budget Estimates, Navy, Justification Book Volume 1 of 5, Research, Development, Test & Evaluation, Navy Budget Activities 1,2, and 3*, May 2017.

[25] Department of Defense, *Department of Defense Fiscal Year (FY) 2018 Budget Estimates, Air Force Justification Book Volume 1 of 3, Research, Development, Test & Evaluation, Vol-1*, May 2017.

university-based basic research across a wide range of scientific and engineering disciplines pertinent to maintaining land combat technology superiority. Army MURI efforts involve teams of researchers investigating high-priority, transformational topics that intersect more than one traditional technical discipline. The MURI multidisciplinary approach seeks to accelerate research progress and to expedite the transition of research results into application. The DURIP program provides funds to acquire major research equipment to augment current research capabilities, or to devise new research capabilities, in support of Army transformational research. The PECASE program funds single-investigator research efforts performed by academic scientists and engineers early in their research careers. The Minerva Research Initiative is a university-based social science research program that seeks to improve DOD's basic understanding of the social, cultural, behavioral, and political forces that shape regions of the world of strategic importance to the United States. The program has three primary components: (1) a university-based social science basic research grant program; (2) the Research for Defense Education Faculty program for the professional military education institutions; and (3) a collaboration with the U.S. Institute of Peace to award research support to advanced graduate students and early career scholars working on security and peace.[26] According to DOD:

> The Minerva Research Initiative has a unique relationship between research and policy within DOD. As such, leadership across the department collaborate to identify and support basic social science research issues in need of attention and to integrate those research insights into the policy-making environment. In doing this, the leadership team closely works with the program managers within the Military Service Branches.[27]

---

[26] Department of Defense, *Department of Defense Fiscal Year (FY) 2018 Budget Estimates, Army Justification Book of Research, Development, Test & Evaluation, Army RDT&E, Volume I, Budget Activity 1*, May 2017.

[27] Department of Defense, website, "The Minerva Research Initiative: Supporting Social Science for a Safer World," http://minerva.defense.gov/Minerva.

Navy

The Navy's URI PE includes support for multidisciplinary basic research in a wide range of scientific and engineering disciplines to enable the U.S. Navy to maintain technological superiority, and for the acquisition of research instrumentation needed to maintain and improve the quality of university research important to the Navy. Navy MURI efforts are focused on high priority topics and opportunities that intersect more than one traditional technical discipline. This program is intended to stimulate innovation, accelerate research progress, and expedite transition of research results into naval applications. The Navy DURIP program supports university research infrastructure deemed essential to high quality, Navy-relevant research. The program complements other Navy research programs by supporting the purchase of high-cost research instrumentation that is necessary for the conduct of cutting-edge research. Navy URI funding also supports PECASE efforts focused on providing the knowledge base, scientific concepts, and technological advances needed for the maintenance of naval power and national security.[28]

Air Force

The Air Force's URI PE supports defense-related basic research across a wide range of scientific and engineering disciplines deemed relevant to maintaining U.S. military technological superiority. Research topics include transformational and high priority technologies such as: nanotechnology, sensor networks, intelligence information fusion, smart materials and structures, efficient energy and power conversion, and high-energy materials for propulsion and control. The program also seeks to enhance and promote the education of U.S. scientists and engineers in disciplines critical to maintaining, advancing, and enabling future U.S. defense technologies. The program also assists universities in acquiring the instrumentation capabilities needed to improve the quality of defense-related research and education. The Air Force asserts that a fundamental

---

[28] Department of Defense, *Department of Defense Fiscal Year (FY) 2018 Budget Estimates, Navy, Justification Book Volume 1 of 5, Research, Development, Test and Evaluation, Navy Budget Activities 1,2, and 3*, May 2017.

component of this program is recognition that future technologies and technology exploitations require highly coordinated and concerted multi- and interdisciplinary efforts.[29]

In-House Laboratory Independent Research

The In-House Laboratory Independent Research (ILIR) PEs support basic research at Army and Navy laboratories, including:

- six Army Materiel Command Research, Development, and Engineering Centers; six U.S. Army Medical Research and Materiel Command Laboratories; seven Corps of Engineers U.S. Army Engineer Research and Development Centers; the U.S. Space and Missile Defense Command Technical Center; and
- participating Naval Warfare Centers and Laboratories.

Army

Army ILIR efforts seek to catalyze major technology breakthroughs by providing laboratory directors flexibility in implementing novel research ideas, by nurturing promising young scientists and engineers, and by attracting and retaining top scientists and engineers. The ILIR program also provides a source of competitive funds for peer reviewed efforts at Army laboratories to stimulate high quality, innovative research with significant opportunity for payoff to Army warfighting capability.[30]

Navy

Navy ILIR efforts are selected by Naval Warfare Centers/Laboratories' commanding officers and technical directors near the start of each fiscal year through internal competition. Efforts typically last three years, and are generally designed to assess the promise of new lines of research.

---

[29] Department of Defense, *Department of Defense Fiscal Year (FY) 2018 Budget Estimates, Air Force Justification Book Volume 1 of 3, Research, Development, Test and Evaluation, Vol-1*, May 2017.

[30] Department of Defense, *Department of Defense Fiscal Year (FY) 2018 Budget Estimates, Army Justification Book of Research, Development, Test and Evaluation, Army RDT&E, Volume I, Budget Activity 1*, May 2017.

Successful efforts typically attract external, competitively awarded funding.[31]

University and Industry Research Centers

The Army University and Industry Research Centers PE seeks to foster university- and industry-based research to provide a scientific foundation for enabling technologies for future force capabilities. The work falls broadly into three categories:

Collaborative Technology Alliances/Collaborative Research Alliances (CTAs/CRAs)

CTAs seek to leverage large investments by the commercial sector in basic research areas that are of interest to the Army. CTAs are industry-led partnerships between industry, academia, and the Army Research Laboratory (ARL) that seek "to incorporate the practicality of industry, the expansion of the boundaries of knowledge from universities, and Army scientists to shape, mature, and transition technology relevant to the Army mission."[32] CRAs are academia-led partnerships, which seek to leverage cutting-edge academic research.

University Centers of Excellence (COEs)

University COEs seek to expand the frontiers of knowledge in research areas where the Army has enduring needs. COEs couple state-of-the-art research programs at academic institutions with broad-based graduate education programs to help increase the supply of scientists and engineers in automotive and rotary wing technology.

The Army University and Industry Research Centers program element also supports the Historically Black Colleges and Universities and Minority Institution (HBCU/MI) Centers of Excellence.

---

[31] Department of Defense, *Department of Defense Fiscal Year (FY) 2018 Budget Estimates, Navy, Justification Book Volume 1 of 5, Research, Development, Test and Evaluation, Navy Budget Activities 1,2, and 3*, May 2017.

[32] Department of Defense, *Department of Defense Fiscal Year (FY) 2018 Budget Estimates, Air Force Justification Book Volume 1 of 3, Research, Development, Test and Evaluation, Vol-1*, May 2017.

University Affiliated Research Centers (UARCs)

UARCs were established to advance new capabilities through sustained multidisciplinary efforts. The Institute for Soldier Nanotechnologies focuses on soldier protection by emphasizing revolutionary materials research for advanced soldier protection and survivability. The Institute for Collaborative Biotechnologies focuses on enabling network centric-technologies, and broadening the Army's use of biotechnology for the development of bio-inspired materials, sensors, and information processing. The Institute for Creative Technologies is a partnership with academia and the entertainment and gaming industries to leverage innovative research and concepts for training and simulation, in areas such as realistic immersion in synthetic environments, networked simulation, standards for interoperability, and tools for creating simulated environments.

Other DOD Basic Research PEs

In addition to the program elements discussed above, there are seven other DOD basic research program elements (sponsoring agency noted after program name/abbreviation):

- High Energy Laser (HEL) Research Initiatives (Air Force): This PE supports basic research aimed at developing fundamental scientific knowledge to support future DOD HEL systems. This program funds multi-disciplinary research institutes to conduct research on laser and beam control technologies. In addition, this program supports educational grants to stimulate student interest in HELs.
- National Defense Education Program (NDEP, Defense-Wide: Office of the Secretary of Defense): The NDEP supports a number of specific workforce development programs, including the Science, Mathematics, and Research for Transformation program, and the Military Child Pilot Program that seek to improve the DOD workforce by: increasing STEM proficiency in the nation's talent pool; shaping DOD as a STEM workplace of choice for

scientists and engineers through public communications and outreach; leading the DOD STEM strategic efforts and coordinating STEM efforts in alignment with DOD workforce and mission requirements; and identifying approaches for innovative solutions in support of U.S. current and future defense challenges.[33]

- Basic Research Initiatives (D-W: OSD). The Basic Research Initiatives PE supports defense basic research through several activities. Strategic Support for Basic Research (SSBR) initiatives drive the direction of DOD basic research investments; coordinate and conduct oversight of DOD basic research programs; improve science and engineering workforce and public outreach; enhance university-industry collaboration; and engage with academic research community and international partners. The PE also supports the Minerva Research Initiative (discussed above) and the Vannevar Bush Faculty Fellowship Program, which supports research across a broad set of emerging scientific areas with transformative potential.
- Basic Operational Medical Research Science (D-W: DARPA). This PE supports basic research in medical-related information and technology leading to fundamental discoveries, tools, and applications critical to solving defense-related challenges. Efforts focus on identified medical gaps in warfighter care related to health monitoring and preventing the spread of infectious disease. The program uses information, computational modeling, and physical sciences to discover properties of biological systems that cross multiple scales of biological architecture and function, from the molecular and genetic level through cellular, tissue, organ, and whole organism levels. To enable in-theater, continuous analysis and treatment of warfighters, this project seeks to explore diagnostic and therapeutic approaches, including the use of bacterial predators as therapeutics against infections caused by

[33] STEM is an acronym for science, technology, engineering, and mathematics.

antibiotic-resistant pathogens; developing techniques to enable rapid transient immunity for emerging pathogens; and identifying fundamental biological mechanisms that enable certain species to survive in harsh environments.

- Chemical and Biological Defense Program (D-W: Nuclear, Chemical, and Biological Defense Program). The Chemical and Biological Defense Program (CBDP) includes PEs in all seven RDT&E budget activities. The basic research focused Chemical and Biological Defense Program PE supports theoretical and experimental research in life sciences—focused on understanding living systems' response to biological or chemical agents, to support detection, diagnostics, protection, and medical treatment—and physical sciences—focused on investigation of physical and chemical properties and interactions to improve detection, diagnostics, protection, and decontamination.
- Defense Threat Reduction Agency (DTRA) University Strategic Partnership Basic Research (D-W: DTRA). The DTRA Basic Research PE funds research across physical, material, engineering, computational, and life sciences directed toward greater knowledge and understanding of the fundamental aspects of observable phenomena associated with weapons of mass destruction. This PE provides support for the discovery and development of basic knowledge by researchers in academia and research institutions in government and industry.
- Historically Black Colleges and Universities and Minority-Serving Institutions (HBCU/MI, D-W: OSD). The HBCU/MI PE provides support for Historically Black Colleges and Universities and Minority-Serving Institutions (HBCU/MI) programs in the fields of science and engineering deemed important to national defense. This PE provides support through grants, cooperative agreements, or contracts for research, education assistance, and instrumentation.

Issues in Defense S&T

Through the authorization and appropriations processes, Congress grapples with a wide variety of issues related to the magnitude, allocation, and strategic direction of DOD RDT&E, Defense S&T (a subset of RDT&E), and basic research (a subset of Defense S&T). These decisions play an important role in U.S. national security and economic strength, in the near term and longer term.

In practice, appropriations decisions are generally made about specific programs within the context of the available funding. The levels of RDT&E, S&T, and basic research funding are the result of many decisions made during DOD budget formulation and congressional appropriations, and in the end, are calculated on a post-facto basis. Nevertheless, an analysis of the kind that follows may be useful in assessing the big picture and in seeing funding trends in the context of a historical arc that may provide strategic insight and guidance.

Among the ongoing questions lawmakers and policy analysts grapple with are:

- What is the appropriate funding level for Defense S&T?
- What is the appropriate funding level for DOD basic research?

Several approaches to addressing these questions are identified below, each with related data and analysis.[34]

---

[34] A question related to Defense S&T funding, is "What is the appropriate funding level for DOD research targeting revolutionary technological advancements?" However, DOD does not request, receive, or report funding using a taxonomy that includes "revolutionary research." In the absence of such an accounting, funding for the Defense Advanced Projects Research Agency (DARPA) has served as a surrogate measure of such DOD research. For a more detailed analysis of DARPA and its appropriations, see CRS Report R45088, *Defense Advanced Research Projects Agency: Overview and Issues for Congress*, by Marcy E. Gallo.

What Is the Appropriate Funding Level for Defense S&T?

Congress and others have expressed concerns about the adequacy of funding for Defense S&T. As discussed earlier, the scientific and technological insights that emerge from this funding are seen by many as the pool of knowledge available to DOD and the industrial base for future defense technology development. For this reason, Defense S&T funding has sometimes been singled out for attention by Congress.

Approach: Defense S&T as a Share of Total DOD Funding

A 1998 Defense Science Board (DSB) report suggested two conceptual frameworks for Defense S&T funding. The first approach, using industrial practice as a guide, proposed setting Defense S&T funding at 3.4% of total DOD funding:

> The DOD S&T budget corresponds most closely to the research component of industrial R&D. Using 3.4% of revenue (typical of high-tech industries shown [elsewhere in the report]), the DOD S&T funding should be about $8.4 billion, which is a billion dollars greater than the FY98 S&T funding.[35]

Other organizations have proposed using the same metric, with 3% of total DOD funding as the level for S&T funding. A 2001 report based on the Quadrennial Defense Review (QDR), a legislatively mandated review by DOD of its strategies and priorities, called for "a significant increase in funding for S&T programs to a level of three percent of DOD spending per year."[36] In 2004, the Council on Competitiveness, a leadership organization of corporate chief executive officers, university presidents, labor leaders, and national laboratory directors, reiterated the 3% recommendation of the QDR.[37]

---

[35] Defense Science Board, *Report of the Defense Science Board Task Force on Defense Science and Technology Base for the 21st Century*, June 1998.

[36] Department of Defense, *Quadrennial Defense Review Report*, September 30, 2001, p. 41, http://archive.defense.gov/ pubs/qdr2001.pdf.

[37] Council on Competitiveness, *Innovate America*, 2004, p. 58, http://www.compete.org/ storage/images/uploads/File/PDF%20Files/NII_Innovate_America.pdf.

Over the years, Congress has sought to address this perceived shortcoming in funding. The FY1999 defense authorization bill (P.L. 105-261) expressed the sense of Congress that Defense S&T funding should be increased by 2% or more above the inflation rate each year from FY2000 to FY2008.[38] Subsequently, the FY2000 defense authorization bill expressed the sense of Congress that

> the Secretary of Defense has failed to comply with the funding objective for the Defense Science and Technology Program, especially the Air Force Science and Technology Program, as stated [P.L. 105-261], thus jeopardizing the stability of the defense technology base and increasing the risk of failure to maintain technological superiority in future weapon systems.[39]

The act further expressed the sense of Congress that the Secretary of Defense should increase Defense S&T, including the 6.1-6.3 programs within each military department, by 2% or more above the inflation rate each year from FY2001 to FY2009.

In 2002, Congress embraced the DSB's recommendation and underlying rationale in the conference report accompanying the National Defense Authorization Act for Fiscal Year 2003:

> The conferees commend the Department of Defense commitment to a goal of three percent of the budget request for the defense science and technology program and progress toward this goal. The conferees also note the finding in the Defense Science Board report that successful high technology industries invest about 3.5 percent of sales in research (equivalent to the DOD S&T program) and the recommendation that S&T funding should be increased to ensure the continued long-term technical superiority of U.S. military forces in the 21st Century. The conferees believe that the Department must continue to provide the necessary investments in research and technologies that ensure a strong, stable, and robust science and technology program for our Armed Forces.[40]

---

[38] P.L. 105-261, Sec. 214.

[39] P.L. 106-65.

[40] H.Rept. 107-772, p. 460, http://lis.gov/cgi-lis/t2gpo/https:/www.gpo.gov/fdsys/pkg/CRPT-107hrpt772/pdf/CRPT107hrpt772.pdf.

In 2009, the Senate-passed version of the National Defense Authorization Act for Fiscal Year 2010 (S. 1390) included a provision (Sec. 217) stating it was the sense of Congress that the Secretary of Defense should increase Defense S&T by a percentage at least equal to inflation.

Data and Analysis

Following a period of strong growth in the early 2000s, Defense S&T funding reached $13.3 billion in FY2006, then declined to $11.0 billion in FY2013 before climbing to a peak of $14.0 billion in FY2017 (See Figure 6). Growth in the amount of S&T funding that was sought in P.L. 105-261 (red line, Figure 6) was largely achieved, though appropriations fell somewhat short in FY2007 and FY2008. Viewed as a share of DOD total obligational authority (TOA), S&T declined from about 3.0% in the late 1990s to about 1.7% in 2011, then rebounded to about 2.3% in FY2017 (See Figure 7).

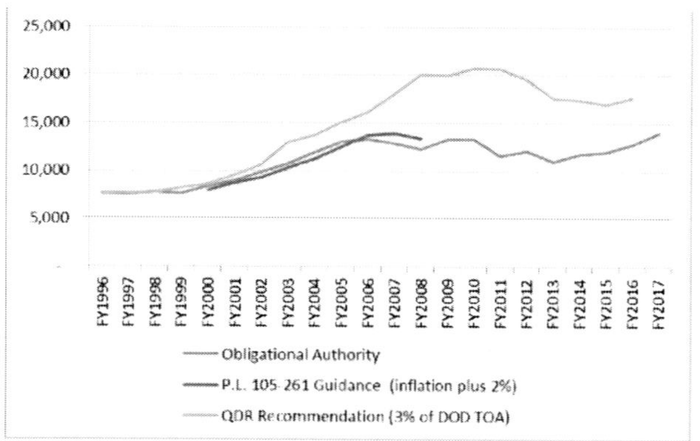

Source: CRS analysis of data from Department of Defense, *Research, Development, Test, and Evaluation Programs (R-1)* for FY1998-2019.

Note: For purposes of this chart, CRS used the GDP (Chained) Price Index from Table 10.1 of the Historical Tables in the *President's Budget for Fiscal Year 2018* to adjust for inflation. This is the index used by the Office of Management and Budget to convert federal research and development outlays from current dollars to constant dollars. https://www.whitehouse.gov/sites/default/files /omb/budget/fy2017/assets/hist10z1.xls.

Figure 6. Defense S&T Funding, P.L. 105-261 Guidance, and QDR Recommendation; in millions of current dollars.

Approach: DOD Science and Technology as a Share of DOD RDT&E

The DSB's second proposed framework, also based on industrial practice, was to use the metric of Defense S&T as a share of DOD RDT&E:

> Another approach to this question is to note that the ratio of research funding to total R&D funding in high-technology industries, such as pharmaceuticals, is about 24%. When this percentage ratio is applied to the FY98 R&D funding of about $36 billion, the result is about $8.6 billion, well above the actual S&T funding.[41]

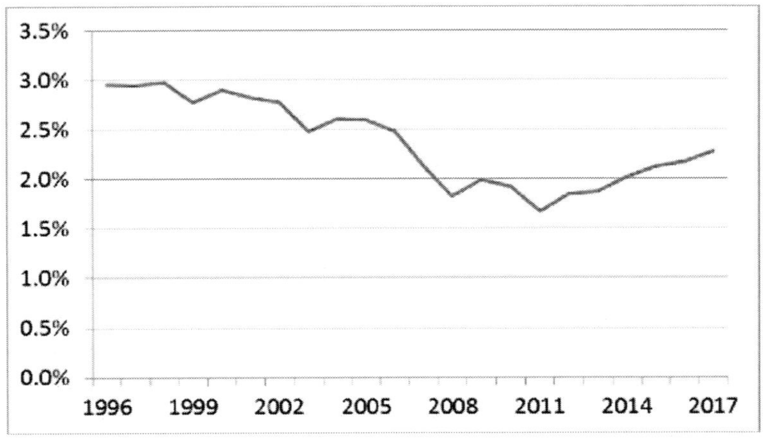

Source: CRS analysis of data from Department of Defense, Research, Development, Test, and Evaluation Programs (R-1) for FY1998-2019; DOD, National Defense Budget Estimates for FY2018 (Green Book), FY2017.

Note: TOA=Total Obligational Authority.

Figure 7. Defense S&T Funding as a Share of DOD TOA.

---

[41] Defense Science Board, *Report of the Defense Science Board Task Force on Defense Science and Technology Base for the 21st Century*, June 1998. Some analysts may disagree with DSB's implicit assumption about the applicability of a ratio drawn from the R&D investment behavior of private firms competing in a commercial market to DOD S&T spending.

In 2015, the Coalition for National Security Research, a coalition of industry, universities, and associations, asserted that Defense S&T funding should be 20% of DOD RDT&E.[42]

Data and Analysis

Figure 8 illustrates Defense S&T's share of DOD RDT&E for FY1996-FY2016. At the time of the DSB report (1998), S&T's share of DOD RDT&E was approximately 20.7%. After rising to 21.5% in FY2000, the share fell to 15.2% in FY2011, and then recovered to 17.9% in FY2017.

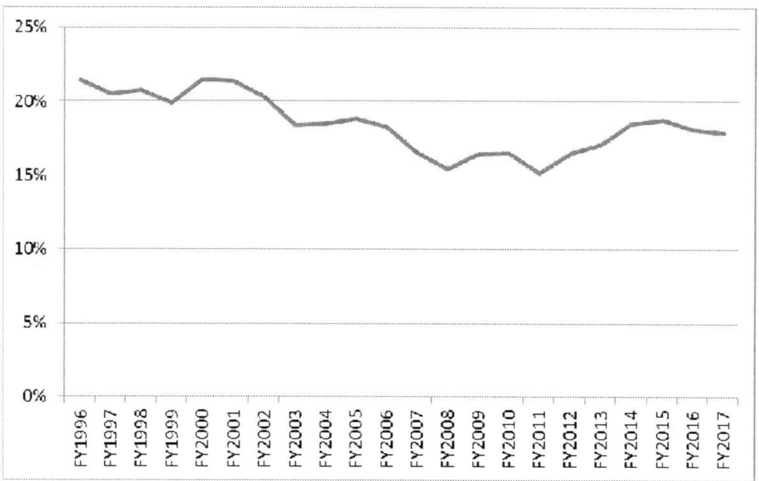

Source: CRS analysis of data from Department of Defense, *Research, Development, Test, and Evaluation Programs (R-1)* for FY1978-2019. CRS used funding levels from two years before the request year. For example, the FY2017 funding levels are from the FY2019 R-1.

Note: Congress appropriates some RDT&E funding in other appropriations titles. In 2016, for example, Congress appropriated $2.121 billion for the Defense Health Program, $579 million for Chemical Agents and Munitions Destruction, $26 million for the National Defense Sealift Fund, and $2.1 million for the Inspector General for R&D and RDT&E-related purposes. In total, these funds accounted for 3.7% of all RDT&E funds.

Figure 8. Defense S&T Funding as a Share of DOD Title IV RDT&E.

---

[42] Richard M. Jones, "Coalition Recommends Higher Level of Defense S&T Funding than Administration Request," *FYI: Science Policy News from AIP*, April 13, 2015, https://www.aip.org/fyi/2015/coalition-recommends-higher-leveldefense-st-funding-administration-request.

What Is the Appropriate Funding Level for DOD Basic Research?

Within the Defense S&T program, basic research is often singled out for additional attention, due in part to its perceived value in advancing breakthrough technologies and in part to the substantial role it plays in supporting university-based research in certain physical sciences and engineering disciplines. DOD describes basic research as "farsighted high payoff research that provides the basis for technological progress."[43] Basic research funding is seen by some as particularly vulnerable to budget cuts or reallocation to other priorities because of the generally long time it takes for basic research investments to result in tangible products and other outcomes (i.e., reductions in funding can be made with minimal short-term consequences) and to the uncertainty of the benefits that will be derived from the results of basic research.

Approach: DOD Basic Research as a Share of Defense S&T

In 2004, the Council on Competitiveness asserted that DOD basic research should be at least 20% of Defense S&T.[44] In 2015, the Coalition for National Security Research also recommended basic research account for 20% of Defense S&T.[45]

Data and Analysis

DOD basic research funding grew steadily from FY1998 through FY2017, more than doubling in current dollars (See Figure 9). As a share of Defense S&T, basic research declined from 14.6% in FY1996 to 11.0% in FY2006, then began a steady rise to 18.4% in FY2015, its highest level in two decades. Basic research's share of Defense S&T fell in 2016 to 17.4% and in 2017 to 16.2% (See Figure 10).

---

[43] Department of Defense, *Financial Management Regulation (DoD 7000.14-R),* Volume 2B, March 2016.

[44] Council on Competitiveness, *Innovate America*, 2004, p. 58.

[45] Richard M. Jones, "Coalition Recommends Higher Level of Defense S&T Funding than Administration Request," *FYI: Science Policy News from AIP*, April 13, 2015.

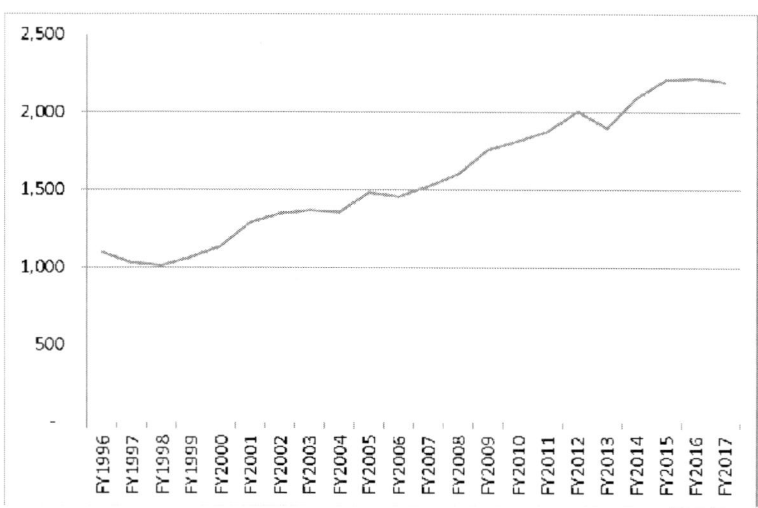

Source: CRS analysis of data from Department of Defense, *Research, Development, Test, and Evaluation Programs (R-1)* for FY1998-2019.

Figure 9. DOD Basic Research Funding; In millions of current dollars.

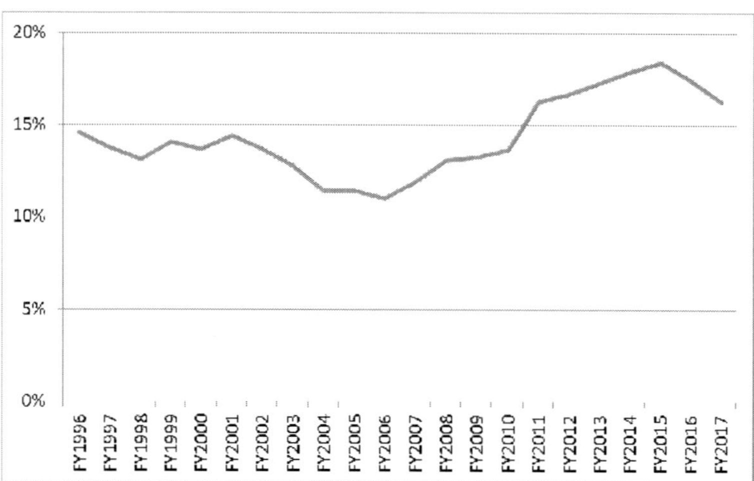

Source: CRS analysis of data from Department of Defense, *Research, Development, Test, and Evaluation Programs (R-1)* for FY1998-2019.

Figure 10. DOD Basic Research as a Share of Defense S&T.

CONCLUSION

Defense S&T investments are highly complex and can be parsed in many ways. Some of these are highlighted in this chapter. Other ways of parsing RDT&E funding—such as allocation by size of industrial performers—may also be important for assessing the balance in allocation of DOD RDT&E resources to meet DOD objectives.

Among the many other factors that may affect the effectiveness of the performance of Defense S&T are: organizational structures and relationships; management; workforce recruitment, training and retention; and policies related to cooperative research and technology transfer. Defense S&T stakeholders have also asserted the importance of stability in funding streams.

As Congress undertakes defense annual authorization and appropriations, it may wish to consider the issues raised in this chapter related to the magnitude and composition of funding for Defense S&T in the overall context of DOD RDT&E, as well as the other issues such as those identified above.

In: Science Policies and Programs
Editor: Johnnie Rodgers
ISBN: 978-1-53614-107-8

Chapter 3

THE NATIONAL SCIENCE FOUNDATION: FY2018 APPROPRIATIONS AND FUNDING HISTORY*

Laurie A. Harris

ABSTRACT

The National Science Foundation (NSF) supports basic research and education in the non-medical sciences and engineering. NSF is a major source of federal support for U.S. university research, especially in certain fields such as computer science. It is also responsible for significant shares of the federal science, technology, engineering, and mathematics (STEM) education program portfolio and federal STEM student aid and support.

Overall, the Trump Administration is seeking $6.653 billion for NSF in FY2018, an $819 million decrease (-11%) from the FY2017 enacted level of $7.472 billion. NSF has six appropriations accounts: Research and Related Activities (RRA), Education and Human Resources (EHR), Major Research Equipment and Facilities Construction (MREFC),

* This is an edited, reformatted, and augmented version of a Congressional Research Service report, R45009, dated December 18, 2017.

Agency Operations and Award Management (AOAM), National Science Board (NSB), and Office of Inspector General (OIG). The FY2018 request would decrease total budget authority primarily in three accounts relative to FY2017 enacted funding: RRA by $672 million (-11%), EHR by $119 million (-14%), and MREFC by $26 million (-12%). The request would provide slight decreases to AOAM ($1.5 million decrease, -0.5%) and OIG ($200,000 decrease, -1.3%), and no change for NSB.

As reported by the House Committee on Appropriations, H.R. 3267 would provide a total of $7.340 billion to NSF for FY2018. This amount is $133 million below (-1.8%) the FY2017 enacted funding level and $687 million (10.3%) above President Trump's FY2018 request. The bill would keep funding for the RRA, EHR, NSB, and OIG accounts the same as the FY2017 enacted amounts and decrease the MREFC and AOAM accounts by $131 million (-62.8%) and $1.5 million (-0.5%), respectively. The text of H.R. 3267 was incorporated into the omnibus appropriations bill, the Make America Secure and Prosperous Appropriations Act, 2018 (H.R. 3354, Division C), and passed, as amended, by the House on September 14, 2017. H.R. 3354 would provide the same total funding amounts for NSF accounts as provided in H.R. 3267.

As reported by the Senate Committee on Appropriations, S. 1662 would provide a total of $7.311 billion to NSF for FY2018. This amount is $161 million below (-2.2%) the FY2017 enacted funding level, and $658 million above (9.9%) President Trump's FY2018 funding request. Compared to the FY2017 enacted level, this bill would keep funding for the NSB and OIG accounts the same and decrease funding for four accounts: RRA by $116 million (-1.9%), MREFC by $26.2 million (-12.5%), EHR by $17.6 million (-2%), and AOAM by $1.5 million (-0.5%).

The Continuing Appropriations Act, 2018 (P.L. 115-56, Division D), signed by the President on September 8, 2017, provided funding for NSF through December 8, 2017, at the FY2017 level, subject to a 0.6791% across-the-board decrease. The Further Continuing Appropriations Act, 2018 (P.L. 115-90, Division A) amends P.L. 115-56 to extend funding through December 22, 2017.

Overall growth in the NSF budget has slowed since FY2003. Average annual growth in NSF appropriations was 8% between FY1997 and FY2003, 4% from FY2004 to FY2010, and 1% between FY2011 and FY2017. Among NSF's appropriations accounts, RRA has accounted for the lion's share of growth in obligations since FY2003. Agency appropriations levels were last authorized in FY2010 and expired in FY2013. Various reauthorization measures were introduced in the 114th Congress that included proposed funding levels; none were enacted. In the 115th Congress, the American Innovation Act (H.R. 1569 and S. 641), introduced as companion bills in March 2017, would authorize increasing appropriations for NSF through FY2021 and adjust the discretionary spending limits to accommodate those increases.

INTRODUCTION

The National Science Foundation (NSF) supports basic research and education in the non-medical sciences and engineering. Congress established the foundation through the National Science Foundation Act of 1950 to "promote the progress of science; to advance the national health, prosperity, and welfare; to secure the national defense; and for other purposes." The NSF is a major source of federal support for U.S. university research, especially in certain fields such as computer science. It is also responsible for significant shares of the federal science, technology, engineering, and mathematics (STEM) education program portfolio and federal STEM student aid and support.

This chapter describes selected items from the Trump Administration's FY2018 budget request for NSF and tracks legislative action on FY2018 appropriations to the agency.[1] It also details selected measures proposed in the 115th Congress to authorize increases in NSF appropriations limits and presents information on historical funding for the agency.

NSF has six appropriations accounts: Research and Related Activities (RRA), Education and Human Resources (EHR), Major Research Equipment and Facilities Construction (MREFC), Agency Operations and Award Management (AOAM), National Science Board (NSB), and the Office of the Inspector General (OIG). Appropriations are generally provided at the account level; program-specific direction may be included in appropriations acts or in accompanying conference reports or explanatory statements. At times, authorizations and appropriations have been specified at the RRA subaccount level, and NSF's budget

[1] Appropriations to NSF are typically included in annual Commerce, Justice, Science and Related Agencies Appropriations Acts. The Congressional Research Service tracks these acts on CRS.gov, at http://www.crs.gov/AppropriationsStatusTable/index.

justifications detail activities and obligations at that level.[2] The majority of NSF's primary mission activities are funded through RRA, EHR, and MREFC. NSF adopted its current appropriations account structure in FY2003. In general, NSF's accounts have been comparable since then.[3]

For FY2017, the Consolidated Appropriations Act, 2017 (P.L. 115-31), signed by the President on May 5, 2017, provided appropriations at the account level. Congress also directed funding for a subset of programs within the RRA, EHR, and MREFC accounts in the accompanying explanatory statement.[4] In this chapter, because NSF states that FY2017 amounts were not available when the FY2018 budget request was prepared, requested funding at the account level (and program-specific funding, where directed by Congress) is compared to FY2017 enacted funding. In contrast, requested funding at the subaccount level is generally compared to FY2016 actual levels.[5]

FY2018 Budget and Appropriations Actions

The Trump Administration is seeking $6.653 billion for NSF in FY2018, an $819 million decrease (-11%) from the FY2017 enacted level of $7.472 billion (see Table 1). The request would decrease budget authority primarily in three accounts relative to the FY2017 enacted levels: RRA by $672 million (-11.1%), EHR by $119 million (-13.6%), and MREFC by $26.2 million (-12.5%). The request would provide slight decreases to the AOAM ($1.5 million decrease, -0.5%) and OIG ($200,000

---

[2] NSF's budget justifications are published on the agency's website at http://www.nsf.gov/about/budget/.

[3] In FY2008, NSF shifted the EPSCoR program from the Education and Human Resources (EHR) account to the Research and Related Activities (RRA) account. For more information on EPSCoR, see CRS Report R44689, *Established Program to Stimulate Competitive Research (EPSCoR): Background and Selected Issues*, by Laurie A. Harris.

[4] Explanatory Statement, Consolidated Appropriations Act, 2017, Division B (Commerce, Justice, Science, and Related Agencies Appropriations Act, 2017), *Congressional Record*, vol. 163, no. 76—Book II (May 3, 2017), p. H3375.

[5] Long-term, multi-year construction projects supported through the MREFC account are an exception, as NSF is able to provide FY2017 estimated funding amounts for these projects.

decrease, -1.3%) accounts. The NSB account would receive $4.4 million, the same amount as in FY2017.

As reported by the House Committee on Appropriations on July 17, 2107, H.R. 3267, the Commerce, Justice, Science, and Related Agencies Appropriations Act, 2018, would provide a total of $7.340 billion to NSF for FY2018. This would be a decrease of $133 million (-1.8%) from the FY2017 enacted funding level and an increase of $687 million (10.3%) over President Trump's FY2018 request. The bill would keep funding for the RRA, EHR, NSB, and OIG accounts the same as the FY2017 enacted amounts, and decrease the MREFC and AOAM accounts by $131 million (-62.8%) and $1.5 million (-0.5%), respectively.[6] The text of H.R. 3267 was incorporated as Division C into the omnibus appropriations bill, the Make America Secure and Prosperous Appropriations Act, 2018 (H.R. 3354), and passed, as amended, by the House on September 14, 2017. H.R. 3354 would provide the same total funding amounts for NSF accounts as provided in H.R. 3267.[7]

As reported by the Senate Committee on Appropriations on July 27, 2017, S. 1662, the Commerce, Justice, Science, and Related Agencies Appropriations Act, 2018, would provide a total of $7.311 billion to NSF for FY2018. This would be a decrease of $161 million (-2.2%) from the FY2017 enacted funding level and $658 million (9.9%) over President Trump's FY2018 funding request. Compared to the FY2017 enacted level, this bill would keep funding for the NSB and OIG accounts the same and decrease funding for four accounts: RRA by $116 million (-1.9%), MREFC by $26.2 million (-12.5%), EHR by $17.6 million (-2%), and AOAM by $1.5 million (-0.5%).[8]

---

[6] U.S. Congress, House Committee on Appropriations, *Commerce, Justice, Science, and Related Agencies Appropriations Bill, 2018*, report to accompany H.R. 3267, 115th Cong., 1st sess., July 17, 2017, H.Rept. 115-231 (Washington: GPO, 2017), pp. 68-72.

[7] H.Amdt. 382 to H.R. 3354, agreed to by voice vote in the House, aims to shift funding for certain research areas within amounts provided to the RRA account. The amendment is discussed in the "Research and Related Activities (RRA)" section of this report.

[8] U.S. Congress, Senate Committee on Appropriations, *Departments of Commerce and Justice, Science, and Related Agencies Appropriations Bill, 2018*, report to accompany S. 1662, 115th Cong., 1st sess., July 27, 2017, S.Rept. 115-139 (Washington: GPO, 2017), pp. 114-121.

The Continuing Appropriations Act, 2018 (P.L. 115-56, Division D), signed by the President on September 8, 2017, provided funding for NSF at the FY2017 level through December 8, 2017, subject to a 0.6791% across-the-board decrease. The Further Continuing Appropriations Act, 2018 (P.L. 115-90, Division A) amends P.L. 115-56 to extend funding through December 22, 2017.

The FY2018 NSF budget justification highlights many of the same programs as in FY2016 and FY2017. Specifically, NSF identifies seven ongoing agency-wide investments that aim to bring researchers from different fields of science and engineering together to address cross-disciplinary questions. Compared to the FY2016 actual amounts, a slight increase in funding is requested for one of these initiatives—the Inclusion across the Nation of Communities of Learners of Underrepresented Discoverers in Engineering and Science (INCLUDES, $15 million requested, +6.5%). Decreases of between 12% and 70% from FY2016 actual amounts are requested for the remaining six investments. These include:

- Cyber-Enabled Materials, Manufacturing, and Smart Systems (CEMMSS, $222 million requested, -18%);
- Innovations at the Nexus of Food, Energy, and Water Systems (INFEWS, $24 million requested, -70%);
- Innovation Corps (I-Corps, $26 million requested, -12%);
- Risk and Resilience ($31 million requested, -27%);
- Secure and Trustworthy Cyberspace (SaTC, $114 million requested, -12%); and
- Understanding the Brain (UtB), $134 million requested, -22%).

The committee reports to accompany H.R. 3267 (H.Rept. 115-231) and S. 1662 (S.Rept. 115-139) provide direction for one of these programs, recommending no less than the FY2017 level ($30 million) for I-Corps.

NSF's budget request reports that three additional multi-directorate programs highlighted in prior years are ending in FY2017. The Cyberinfrastructure Framework for 21st Century Science, Engineering, and

Education (CIF21) program is sunsetting, though some of the program's activities will be incorporated into NSF's work on the National Strategic Computing Initiative (NSCI) and the Harnessing the Data Revolution Big Idea. The Research at the Interface of Biological, Mathematical, and Physical Sciences (BioMaPS) program is ending "as it has achieved its goal, leading to a cultural change within NSF cross-directorate collaboration [among the participating directorates] having become standard practice." Though funding is concluding for the Science, Engineering, and Education for Sustainability (SEES) program, NSF plans to continue investing in "research necessary for a sustainable human future" through other programs, such as the Risk and Resilience and INFEWS programs.[9]

The budget request also highlights NSF's 10 "Big Ideas," introduced by the agency in 2016 as long-term research and process ideas that identify areas for future investment. The budget request includes various activities related to these ideas, though specific funding is requested only for the INCLUDES program ($14.9 million). Research ideas include:

- Harnessing the Data Revolution;
- Work at the Human Technology Frontier: Shaping the Future;
- Windows on the Universe: The Era of Multi-messenger Astrophysics;
- The Quantum Leap: Leading the Next Quantum Revolution;
- Understanding the Rules of Life: Predicting Phenotype; and
- Navigating the New Arctic.

Process ideas include:

- Mid-scale Research Infrastructure;

---

[9] National Science Foundation, *FY2018 Budget Request to Congress*, May 23, 2017, p. Performance-36. NSF announced $18.7 million in FY2017 awards through the Risk and Resilience program; see the NSF press release, "In Wake of Hurricanes, Floods and Wildfires, NSF Awards $18.7 Million in Natural Hazards Research Grants," September 12, 2017, at https://www.nsf.gov/news/news_summ.jsp?cntn_id=242941.

- NSF 2026: Seeding Innovation;
- NSF INCLUDES; and
- Growing Convergent Research at NSF.[10]

The committee reports do not specify funding for the Big Ideas. Broadly, both reports express support for such topics as scientific infrastructure investments, astronomy facilities, cybersecurity research, and broadening participation of underrepresented groups in STEM. The reports also encourage NSF to explore partnerships with the private sector in supporting various facilities, equipment, and programs.

Table 1. NSF Funding by Account, FY2016-FY2018 (budget authority in millions of dollars)

			FY2018			
Account	FY2016 Actual	FY2017 Enacted	Request	House Committee-Reported	Senate Committee-Reported	Enacted
Research and Related Activities (RRA)	$5,998.1	$6,033.6	$5,361.6	$6,033.6	$5,917.8	
Education and Human Resources (EHR)	884.1	880.0	760.6	880.0	862.4	
Major Research Equipment and Facilities Construction (MREFC)	241.5	209.0	182.8	77.8	182.8	
Agency Operations and Award Management (AOAM)	351.1	330.0	328.5	328.5	328.5	
National Science Board (NSB)	4.3	4.4	4.4	4.4	4.4	
Office of the Inspector General (OIG)	14.8	15.2	15.0	15.2	15.2	
NSF, Total	$7,493.9	$7,472.2	$6,652.9	$7,339.5	$7,311.1	

Source: FY2018 *NSF Budget Request to Congress*; H.R. 3267 as reported by the House Committee on Appropriations on July 17, 2017, and H.Rept. 115-231; H.R. 3354 as passed by the House of Representatives on September 14, 2017; S. 1662, as reported by the Senate Committee on Appropriations on July 27, 2017, and S.Rept. 115-139.

Notes: Totals may not add due to rounding.

---

[10] National Science Foundation, *FY2018 Budget Request to Congress*, May 23, 2017, p. Overview-3. See also National Science Foundation, "10 Big Ideas for Future NSF Investments," at https://www.nsf.gov/about/congress/reports/ nsf_big_ideas.pdf.

Research and Related Activities (RRA)

The Trump Administration is seeking $5.362 billion for RRA in FY2018, a $627 million decrease (-11%) compared to FY2017 enacted funding. Compared to FY2016 actual levels, the FY2018 request includes decreases for all 10 of the RRA subaccounts except for the U.S. Arctic Research Commission (USARC), which would not change (see Table 2). The largest decrease in dollars would go to Mathematical and Physical Sciences (MPS, $129 million decrease, -9.6%), and the largest percentage decrease would go to Integrative Activities (IA, $111 million decrease, -26%). The other subaccounts would receive decreases between 7.1% and 10.6%. The FY2018 request also includes $100 million for the RRA program Established Program to Stimulate Competitive Research (EPSCoR), a decrease of $60 million (-37.5%) from the $160 million directed in the explanatory statement for FY2017 enacted funding.

The House and Senate bills would both provide funding for RRA and EPSCoR at the same, or slightly below, FY2017 enacted funding levels. As reported by the House Committee on Appropriations, H.R. 3267 would provide a total of $6.034 billion for RRA, equal to the FY2017 enacted level. The House committee report states:

> The Committee does not adopt the Administration's proposal to reduce Research and Related Activities. The Committee believes that strategic investments in the physical science areas are vitally important for the United States to remain the global leader in innovation, productivity, economic growth, and good-paying jobs for the future.[11]

As reported by the Senate Committee on Appropriations, S. 1662 would provide a total of $5.918 billion for RRA, a decrease of $115.8 million (-1.9%) from the FY2017 enacted level.

For EPSCoR, the House committee report recommends $170.7 million, $70.7 million (71%) more than the FY2018 request and $10.7 million (7%) above the FY2017 enacted level. The Senate committee report

---

[11] H.Rept. 115-231, p. 69.

recommends no less than $160 million for EPSCoR, equal to the FY2017 level, and directs NSF to make efforts to ensure that no more than 5% of program funding is used for administrative and overhead costs.

Neither report specifies funding amounts at the RRA subaccount levels, a practice that has been a point of congressional debate in prior years, though the reports do specify funding levels for certain divisions and programs within RRA directorates. The House report directs the Division of Astronomical Sciences, located in the Directorate for Mathematical and Physical Sciences, to support programs and scientific facilities at no less than the FY2017 levels. Additionally, the House committee report directs $48 million (the requested amount) for the International Ocean Discovery Program, within the Geosciences Directorate. Further, H.Amdt. 382 to Division C of the omnibus appropriations bill (H.R. 3354), agreed to by voice vote in the House, supports physical science funding. While not changing the total funding amount for the RRA account, the amendment's purpose is to "increase physical and biological science research by one-half of one percent, or $30.2 million, over the current funding within [RRA]. Total spending is not increased, as NSF will adjust other areas of spending accordingly."[12]

As in FY2017, Senate committee report language for FY2018 specifies funding levels for the Historically Black Colleges and Universities Excellence in Research program ($10 million) and the Advancement of Women in Academic Science and Engineering Careers program (ADVANCE, $18 million). The House and Senate committee reports both contain language supporting the inclusion of national security and economic competitiveness criteria as part of the merit review processes for research grant proposals. Both reports also encourage NSF to examine its funding portfolio to ensure adequate support of fire research.

[12] Remarks by Rep. Lamar Smith during House debate, *Congressional Record*, daily edition, vol. 163, no. 147 (September 12, 2017), p. H7257.

Table 2. NSF Funding by RRA Subaccount, FY2016-FY2018 (budget authority in millions of dollars)

Account	FY2016 Actual	FY2017 Enacted	FY2018			
			Request	House Committee-Reported	Senate Committee-Reported	Enacted
Biological Sciences (BIO)	$723.8	n/s	$672.1	n/s	n/s	
Computer and Information Science and Engineering (CISE)	935.2	n/s	838.9	n/s	n/s	
Engineering (ENG)	915.7	n/s	833.5	n/s	n/s	
Geosciences (GEO)	876.5	n/s	783.3	n/s	n/s	
Mathematical and Physical Sciences (MPS)	1348.8	n/s	1219.4	n/s	n/s	
Social, Behavioral, and Economic Sciences (SBE)	272.2	n/s	244.0	n/s	n/s	
Office of International Science and Engineering (OISE)	49.1	n/s	44.0	n/s	n/s	
Office of Polar Programs (OPP)	448.9	n/s	409.2	n/s	n/s	
International and Integrative Activities (IIA)	246.6	n/s	315.7	n/s	n/s	
U.S. Arctic Research Commission (USARC)	1.4	n/s	1.4	n/s	n/s	
Research and Related Activities (RRA), Total	$5,998.1	$6,033.6	$5,361.6	$6,079.4	$6,033.6	

Source: FY2018 *NSF Budget Request to Congress*; H.R. 3267 as reported by the House Committee on Appropriations on July 17, 2017, and H.Rept. 115-231; H.R. 3354 as passed by the House of Representatives on September 14, 2017; S. 1662, as reported by the Senate Committee on Appropriations on July 27, 2017, and S.Rept. 115-139.

Notes: The term "n/s" means "not specified." Totals may not add due to rounding.

Education and Human Resources (EHR)

The FY2018 budget request includes $761 million for EHR, a $119 million decrease (-13.6%) from the FY2016 estimate. This represents the largest percentage reduction requested among NSF's appropriations accounts. As reported by the House Committee on Appropriations, H.R. 3267 would provide $880 million for EHR, equal to the FY2017 enacted level and $119.4 million (16%) above the FY2018 request. As reported by the Senate Committee on Appropriations, S. 1662 would provide $862.4 million for EHR, a decrease of $17.6 million (-2%) from the FY2017 enacted level and $101.8 million (13.4%) above the FY2018 request.

Within EHR, there are four divisions: Division of Graduate Education (DGE), Division of Undergraduate Education (DUE), Division of Human Resource Development (HRD), and Division of Research on Learning in Formal and Informal Settings (DRL). By program division, the largest decrease in the FY2018 request is for DGE ($57 million decrease, -20.5%), which would receive $221 million. HRD, DRL, and DUE would receive decreases of 9.4% ($200 million requested), 11% ($135 million requested), and 12% ($204 million requested), respectively.

Programs of particular interest to congressional policymakers within EHR include the Graduate Research Fellowship (GRF) and those with a focus or emphasis on broadening participation among underrepresented minorities in STEM.[13] The FY2018 request for GRF is $246 million, a reduction of $86 million (-26%) from the FY2016 actual level. The requested amount would support 1,000 new fellows, a reduction from the 2,000 new fellows supported through the GRF each year since 2011.[14] Across EHR divisions, NSF's FY2018 budget justification classifies programs as one of three core research areas (also called themes): learning and learning environments, broadening participation and institutional capacity, and STEM professional workforce.[15] Total requested funding for EHR programs as part of the broadening participation and institutional capacity theme for FY2018 is $196 million, $12 million less (-5.7%) than in FY2016. Specifically, the budget request includes:

---

[13] While NSF includes broadening participation as a review criterion for all funding proposals, certain program solicitations and announcements include additional, specific requirements or encouraging language. NSF categorizes these programs as part of its "broadening participation portfolio," within which programs are classified as focused programs, emphasis programs, or geographic diversity programs. See NSF's budget justification, p. Summary Tables-11, for more information.

[14] In addition to new fellows, funding would continue for an estimated 5,000 active fellows.

[15] For additional information on this categorization structure for EHR programs, see NSF Federal Advisory Committee for Education and Human Resources, *Strategic Re-envisioning for the Education and Human Resources Directorate*, May 1, 2014, at http://www.nsf.gov/ehr/Pubs/AC_ReEnvisioning_Report_Sept_2014_01.pdf.

- Advancing Informal STEM Learning (AISL, $62.5 million requested, no change from the FY2016 amount);
- Science, Technology, Engineering, and Mathematics + Computing Partnerships (STEM+C, $20 million requested, -61.4%);
- Historically Black Colleges and Universities Undergraduate Program (HBCU-UP, $35 million requested, no change);
- Tribal Colleges and University Programs (TCUP, $13 million requested, -7.1%);
- Louis Stokes Alliance for Minority Participation (LSAMP, $41 million requested, -10.9%); and
- Improving Undergraduate STEM Learning: Hispanic Serving Institutions program (IUSE: HSI, $15 million requested, no change).[16]

The House and Senate committee reports recommend $35 million for the HBCU-UP program, $14 million for the TCUP program, $46 million for the LSAMP program, and $15 million for the HSI program. The Senate committee report also recommends $62.5 million for AISL and $52 million for STEM+C, equal to FY2016 funding levels. The Senate committee report further states:

> The Committee does not adopt the proposed funding reductions for the NSF Scholarships in STEM, Robert Noyce Scholarship Program, or the Graduate Research Fellowship and instead provides the fiscal year 2017 funding level for these programs.[17]

NSF's budget justification states that "overall, there are no significant shifts in EHR's priorities between FY2016 and FY2018." One of the ongoing EHR priority programs is CyberCorps: Scholarships for Service, for which NSF requests $40 million, $10 million below (-27%) the

---

[16] The explanatory statement accompanying the Consolidated Appropriation Act, 2017 (P.L. 115-31) directed NSF to provide at least $15 million to establish an HSI program. For FY2018, NSF is requesting $15 million for HSIs through the IUSE program with program support from DUE and HRD (see NSF's budget justification, p. EHR-6).

[17] S.Rept. 115-139, p. 118.

FY2017 directed level. The Senate committee report recommends $55 million for this program (10% increase); the House committee report does not specify an amount.

Major Research Equipment and Facilities Construction (MREFC)

The Major Research Equipment and Facilities Construction (MREFC) account supports large construction projects and scientific instruments.[18] The Trump Administration is seeking just over $183 million for MREFC in FY2018, a decrease of $26 million (-12.5%) from the FY2017 enacted amount. Requested MREFC funding would support three main projects, including continued construction of the Large Synoptic Survey Telescope (LSST, $58 million requested, -13.6% compared to the FY2017 estimate) and the Daniel K. Inouye Solar Telescope (DKIST, $20 million requested, no change). Most of the request ($105 million) would fund the Regional Class Research Vessels (RCRV) program to build ships to support science in U.S. coastal waters. The FY2018 request—prepared in advance of final FY2017 appropriations action by Congress—included support for two ships. Subsequently, in the Consolidated Appropriations Act, 2017, Congress directed NSF to provide $122 million to build three RCRVs. This amounts to $41 million per ship, compared to the FY2018 request of $52.5 million per ship. The budget request notes that the direction from Congress for three RCRVs will impact current and future funding requirements at unspecified amounts.

As reported by the House Committee on Appropriations, H.R. 3267 would provide a total of $77.8 million for MREFC in FY2018, $131 million below (-63%) the FY2017 enacted level, and $105 million below (-57%) the FY2018 request. The House committee report directs $57.8 million for LSST and $20 million for DKIST; no funding is recommended

[18] MREFC funding supports the acquisition, construction, and commissioning phases of major research infrastructure. The RRA account provides funding for the initial planning and design, as well as post-construction operations and maintenance.

for RCRVs. As reported by the Senate Committee on Appropriations, S. 1662 would provide a total of $182.8 million for MREFC in FY2018, $26.2 million below (-12.5%) the FY2017 enacted level, and equal to the FY2018 request. The recommended amount includes funding at the requested levels for DKIST and LSST, and directs planning and construction of three RCRVs. The Senate committee report also encourages the Government Accountability Office (GAO) to "continue its annual review of programs funded within MREFC so that GAO can report to Congress shortly after each annual budget submission and semiannually thereafter on the status of large-scale NSF projects and activities."

Other Accounts and Initiatives

The Trump Administration seeks $328.5 million for the Agency Operations and Award Management (AOAM) account, a $1.5 million decrease (-0.5%) from FY2017 enacted funding. In recent years, AOAM funding has included support for relocation activities of NSF's headquarters from Arlington, VA, to Alexandria, VA. NSF is requesting $1 million in FY2018 for decommissioning the prior headquarters buildings and any unanticipated changes. Both the House and Senate bills recommend the requested amount for AOAM.

The budget request includes $15 million for the Office of Inspector General (OIG), which is a decrease of $200,000 (-1.3%) from the FY2017 enacted level, and $4.4 million for the National Science Board (NSB, no change). The House and Senate bills would both provide funding equal to the FY2017 enacted amounts.

The FY2018 request highlighted funding for NSF activities under three multi-agency initiatives:[19]

- National Nanotechnology Initiative (NNI, $389 million requested, a decrease of $122 million [-24%] from FY2016),[20]

[19] See the National Science and Technology Council (NSTC) activities under the "NSF-Wide Investments" section of the NSF FY2017 budget request, pp. 33, 38, and 42.

- Networking and Information Technology Research and Development (NITRD, $1.062 billion requested, a decrease of $157 million [-12.9%] from FY2016),[21] and
- U.S. Global Change Research Program (USGCRP, $264 million requested, a decrease of $85 million [-25.6%] from FY2016).[22]

The Senate and House committee reports do not provide specific direction for NSF's investments in these initiatives in FY2018.

AUTHORIZATIONS OF APPROPRIATIONS

Authorizations of appropriations for NSF, which were last enacted in the America COMPETES Reauthorization Act of 2010 (Pi. 111-358), expired in FY2013. Various reauthorization measures were introduced in the 114th Congress that included proposed funding levels, but no authorizations of appropriations were enacted. Members of the 115th Congress have introduced measures that would address authorizations of appropriations to NSF.

The American Innovation Act (H.R. 1569 and S. 641), introduced as companion bills in March 2017, would adjust annual discretionary spending limits for federal science agencies conducting basic research, including NSF, in FY2017-FY2021 to allow for specified increases in appropriations and would authorize appropriations at these increased levels. The bills also specify that annual appropriations for NSF through FY2021 be at least the amount appropriated in FY2016. Table 3 shows the FY2013 authorization levels, appropriations to NSF in FY2016 and

---

[20] For more information on the NNI program, see CRS Report RL34401, *The National Nanotechnology Initiative: Overview, Reauthorization, and Appropriations Issues*, by John F. Sargent Jr.

[21] For more information on the NITRD program, see CRS Report RL33586, *The Federal Networking and Information Technology Research and Development Program: Background, Funding, and Activities*, by Patricia Moloney Figliola.

[22] For more information on FY2018 federal R&D funding, including the multi-agency NNI, NITRD, and USGCRP initiatives, see CRS Report R44888, *Federal Research and Development Funding: FY2018*, coordinated by John F. Sargent Jr.

FY2017, FY2018 requested amounts, and proposed authorized maximum funding levels for NSF in FY2018 under selected measures from the 115th Congress.

Table 3. NSF Appropriation Authorizations (in millions of dollars)

Account	FY2013 Authorized	FY2016 Actual	FY2017 Enacted	FY2018 Request	Proposed Authorization Acts (FY2018)[a]
					S. 641 and H.R. 1569
Research and Related Activities (RRA)	$6,637.8	$5,998.1	$6,033.7	$5,361.6	*n/s*
Biological Sciences (BIO)	*n/s*	*723.8*	*n/s*	*672.1*	*n/s*
Computer and Information Science and Engineering (CISE)	*n/s*	*935.2*	*n/s*	*838.9*	*n/s*
Engineering (ENG)	*n/s*	*915.7*	*n/s*	*833.5*	*n/s*
Geosciences (GEO)	*n/s*	*876.5*	*n/s*	*783.3*	*n/s*
Mathematical and Physical Sciences (MPS)	*n/s*	*1348.8*	*n/s*	*1219.4*	*n/s*
Social, Behavioral, and Economic Sciences (SBE)	*n/s*	*272.2*	*n/s*	*244.0*	*n/s*
Office of International Science and Engineering (OISE)	*n/s*	*49.1*	*n/s*	*44.0*	*n/s*
Office of Polar Programs (OPP)	*n/s*	*448.9*		*409.2*	
Integrative Activities (IA)	*n/s*	*426.6*	*n/s*	*315.7*	*n/s*
U.S. Arctic Research Commission (USARC)	*n/s*	*1.4*	*n/s*	*1.4*	*n/s*
Education and Human Resources					
(EHR)	1,041.8	884.1	880.0	760.6	*n/s*
Major Research Equipment and Facilities Construction (MREFC)	236.8	241.5	209.0	182.8	*n/s*
Agency Operations and Award Management (AOAM)	363.7	351.1	330.0	328.5	*n/s*
National Science Board (NSB)	4.9	4.3	4.4	4.4	*n/s*
Office of the Inspector General (OIG)	15.0	14.8	15.2	15.0	*n/s*
NSF, Total	$8,300.0	$7,493.9	$7,472.2	$6,652.9	$8,306.0

Source: America COMPETES Reauthorization Act of 2010 (P.L. 111-358); FY2018 NSF congressional budget justification; S. 641 as introduced on March 15, 2017, and H.R. 1569 as introduced on March 16, 2017.

Notes: The term "n/s" means "not specified." Totals may not add due to rounding. Amounts in the "FY2016 actual" column represent total, actual budgetary resources, including annual appropriations, unobligated balances, transfers, and other adjustments. Italicized account names represent RRA subaccounts.

[a] These acts include proposed discretionary funding adjustments through FY2021. This table only includes proposed maximum funding amount for NSF for FY2018.

NSF FUNDING HISTORY

The following sections provide information on authorizations of appropriations, as well as funding data and trends, since the foundation was established in 1950.

Long-Term Funding Trends

Table 4, Figure 1, and Figure 2 show the trends in NSF authorizations, budget requests, and appropriations since the foundation was first authorized in the early 1950s. Except in FY1957, current and constant dollar actual appropriations to NSF grew rapidly between FY1951 and FY1966. After FY1967, appropriations fluctuated (up some years and down in others) until about FY1988. NSF experienced periods of generally sustained growth in current and constant dollar appropriations between FY1989 and FY1995 and again between FY1998 and FY2003. Since FY2003, growth in the NSF budget has slowed compared to prior years. Average annual growth in NSF appropriations was 8% between FY1997 and FY2003, 4% from FY2004 to FY2010, and 1% between FY2011 and FY2017.

NSF Obligations by Account

Table 5 shows NSF obligations by account since FY2003. Prior years are not comparable due to changes in NSF account structure. Most of the growth in total NSF obligations since FY2003 has accrued to the main research account, RRA, which increased by about $1.890 billion in current dollars (45.6%) between FY2003 and the FY2017 enacted amount. Total NSF obligations increased by about $2.103 billion (39%) during this same period. If funded at requested levels, total NSF appropriations would be the lowest since FY2009 in current dollars, and the lowest since FY2002 in constant (FY2018) dollars (excluding ARRA funding in both cases).

Table 4. NSF Authorizations, Budget Requests, and Appropriations: FY1951-FY2018 (in millions of current and constant [FY2018] dollars)

Fiscal Year	Current ($ millions)			Constant (FY2018 $ millions)		
	Authorization	Request	Appropriation	Authorization	Request	Appropriation
1951	such sums	–	0	such sums	–	2
1952	such sums	14	4	such sums	109	27
1953	such sums	15	5	such sums	115	36
1954	such sums	15	8	such sums	113	60
1955	such sums	14	14	such sums	105	107
1956	such sums	31	53	such sums	226	387
1957	such sums	41	40	such sums	291	281
1958	such sums	65	52	such sums	444	353
1959	such sums	140	138	such sums	941	925
1960	such sums	160	153	such sums	1,063	1,014
1961	such sums	190	176	such sums	1,243	1,150
1962	such sums	210	263	such sums	1,360	1,705
1963	such sums	358	323	such sums	2,290	2,063
1964	such sums	589	353	such sums	3,722	2,230
1965	such sums	488	420	such sums	3,029	2,610
1966	such sums	530	480	such sums	3,222	2,917
1967	such sums	525	481	such sums	3,097	2,838
1968	such sums	526	495	such sums	3,000	2,824
1969	525	500	400	2,863	2,727	2,181
1970	478	500	440	2,472	2,587	2,277
1971	538	513	513	2,649	2,527	2,527
1972	653	622	622	3,068	2,925	2,925
1973	697	653	649	3,140	2,943	2,925
1974	633	583	579	2,663	2,452	2,438
1975	808	672	764	3,080	2,563	2,914
1976	787	755	715	2,807	2,694	2,551
1977	811	802	776	2,697	2,668	2,582
1978	879	944	863	2,742	2,943	2,691
1979	930	934	911	2,683	2,695	2,629
1980	1,002	1,006	992	2,659	2,671	2,633
1981	1,115	1,148	1,025	2,695	2,776	2,478
1982	n/a	1,354	1,039	n/a	3,062	2,350
1983	n/a	1,073	1,094	n/a	2,325	2,370
1984	n/a	1,292	1,341	n/a	2,705	2,806
1985	n/a	1,502	1,502	n/a	3,043	3,043
1986	1,517	1,569	1,524	3,005	3,109	3,019
1987	1,685	1,686	1,623	3,265	3,266	3,145
1988	n/a	1,893	1,717	n/a	3,553	3,223
1989	2,050	2,050	1,923	3,700	3,700	3,470
1995	n/a	3,200	3,264	n/a	4,919	5,017

Table 4. (Continued)

Fiscal Year	Current ($ millions)			Constant (FY2018 $ millions)		
	Authorization	Request	Appropriation	Authorization	Request	Appropriation
1996	n/a	3,360	3,220	n/a	5,071	4,859
1997	n/a	3,325	3,270	n/a	4,931	4,849
1998	3,506	3,367	3,431	5,135	4,932	5,026
1999	3,773	3,773	3,676	5,458	5,458	5,318
2000	3,886	3,921	3,912	5,507	5,557	5,544
2001	n/a	4,572	4,431	n/a	6,328	6,132
2002	n/a	4,473	4,823	n/a	6,091	6,569
2003	5,536	5,036	5,323	7,399	6,730	7,114
2004	6,391	5,481	5,589	8,334	7,148	7,288
2005	7,378	5,745	5,482	9,329	7,264	6,932
2006	8,520	5,605	5,589	10,433	6,864	6,844
2007	9,839	6,020	5,890	11,730	7,177	7,022
2008	6,600	6,429	6,125	7,708	7,509	7,154
2009	7,326	6,854	6,494a	8,458	7,913	7,497a
2010	8,132	7,045	6,873	9,306	8,063	7,865
2011	7,424	7,424	6,806	8,327	8,327	7,634
2012	7,800	7,767	7,033	8,592	8,555	7,747
2013	8,300	7,373	6,884	8,988	7,984	7,455
2014	n/a	7,626	7,172	n/a	8,119	7,636
2015	n/a	7,255	7,344	n/a	7,621	7,715
2016	n/a	7,724	7,463	n/a	7,987	7,718
2017	n/a	7,964	7,472	n/a	8,091	7,591
2018		6,653			6,653	

Source: Funding data in the "Authorization" columns are from selected FY1951 to FY2013 NSF authorization acts. Funding data in the "Request" and "Appropriations" columns are from National Science Foundation, Budget Internet Information System, "NSF Requests and Appropriations History," NSF.gov, August 14, 2017, http://dellweb.bfa.nsf.gov/NSFRqstAppropHist/NSFRequestsandAppropriationsHistory.pdf, and P.L. 115-31. To calculate constant FY2018 dollars, CRS used the Gross Domestic Product (Chained) Price Index found in Office of Management and Budget, *Historical Tables*, Table 10.1, May 16, 2017, available at https://www.whitehouse.gov/ sites/whitehouse.gov/files/omb/budget/fy2018/hist10z1.xls.

Notes: As per communication between CRS and NSF dated March 20, 2014, the "Appropriation" column shows funding provided in annual appropriations acts plus adjustments required in those acts, other laws, committee reports, etc. Adjustments include rescissions, sequestration, funding transfers across NSF accounts, supplemental appropriations (not including American Recovery and Reinvestment Act, P.L. 111-5, funding in FY2009), and other changes. Resulting amounts most closely align with NSF's approved Current Plans. The term "n/a" means "not available." The term "such sums" means "such sums as may be necessary" to carry out agency powers and duties.

[a] FY2009 appropriation amounts do not include American Recovery and Reinvestment Act (ARRA; P.L. 111-5) supplemental funding, which provided an additional $3,002 million to NSF. With ARRA included, total FY2009 appropriations to NSF were $9,496 million in current dollars and $10,791 million in constant (FY2017) dollars.

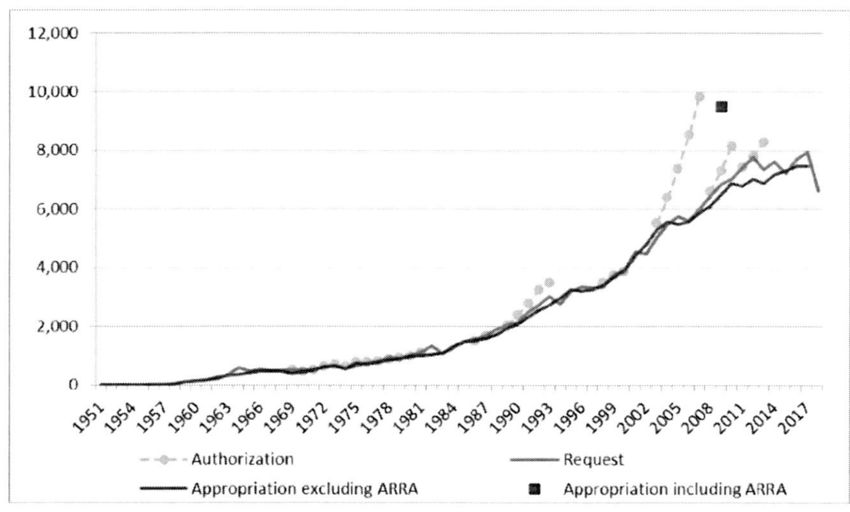

Source: Table 4.

Figure 1. Current Dollar NSF Authorizations, Budget Requests, and Appropriations: FY1951 to FY2018 (in millions of current dollars).

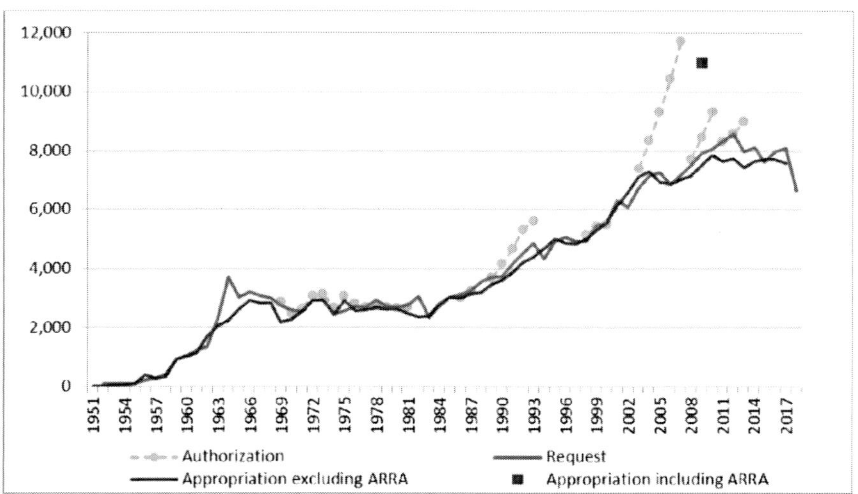

Source: Table 4.

Figure 2. Constant Dollar NSF Authorizations, Budget Requests, and Appropriations: FY1951 to FY2018 (in millions of constant 2018 dollars).

Table 5. NSF Obligations by Account, FY2003-FY2018 Request (in millions of current dollars)

Fiscal Year	RRA	EHR	MREFC	AOAM	NSB	OIG	NSF Total
2003	4,144	846	179	189	3	9	5,369
2004	4,388	850	184	219	2	9	5,652
2005	4,328	750	165	223	4	10	5,481
2006	4,449	700	234	247	4	11	5,646
2007	4,758	696	166	248	4	12	5,884
2008	4,853	766	167	282	4	12	6,084
2009[a]	5,152	846	161	294	4	12	6,469
2010[a]	5,615	873	166	300	4	14	6,972
2011	5,608	861	125	299	4	14	6,913
2012	5,758	831	198	299	4	14	7,105
2013	5,559	835	196	294	4	14	6,902
2014	5,775	832	200	306	4	14	7,131
2015	6,042	886	145	307	4	15	7,398
2016	5,998	884	242	351	4	15	7,494
2017[b]	6,034	880	209	330	4	15	7,472
2018 Request	*5,362*	*761*	*183*	*328*	*4*	*15*	*6,653*

Source: FY2005 to FY2018 annual NSF congressional budget justifications.

Notes: NSF adopted its current appropriations account structure in 2003. For this table, CRS adjusted FY2003 to FY2007 RRA and EHR obligations data to reflect the transfer of the EPSCoR program between these accounts in FY2008. This table treats EPSCoR as part of RRA for all years in the data set.

[a] FY2009 and FY2010 amounts do not include American Recovery and Reinvestment Act (ARRA; P.L. 111-5) supplemental funding. With ARRA included, appropriations in FY2009 were $9,496 million ($3,002 supplemental) total for NSF; $7,686 million ($2,500 million supplemental) for RRA; $945 million ($100 million supplemental) for EHR; $552 million ($400 million supplemental) for MREFC; and $14 million ($2 million supplemental) for OIG. Of the $3,002 million supplemental appropriation for NSF, $2,402 million was obligated in FY2009 and $600 million was obligated in FY2010.

[b] Enacted budget authority, per P.L. 115-31.

POLICY CONSIDERATIONS

To guide decisionmaking for funding reductions in the FY2018 budget request, NSF leadership applied an overarching set of principles, including continuing to fund all science and engineering disciplines, supporting early career scientists, protecting "core" research, and reducing some of the program budgets that have slowly scaled up over the past decade (aka, accretions). Directorates then proposed strategic, prioritized reductions within their program portfolios. Broadly, the request reflects NSF's attempts to maintain an emphasis on cross-disciplinary programs and prioritize programs that will lead to longer-term progress on the Big Ideas.[23]

As in recent years, policymakers are considering congressional funding directives for specific scientific fields within NSF's RRA account. H.Amdt. 382 to Division C of the FY2018 omnibus appropriations bill (H.R. 3354), as well as language in the House committee report accompanying H.R. 3267, emphasize funding support for research in the physical and biological sciences over other scientific fields, such as the social, behavioral, and economic sciences. Supporters of such directives assert that federal dollars should be spent on disciplines they perceive to be more closely tied to research in the national interest (e.g., national security or health) and that such direction falls within Congress's oversight role. Opponents argue that scientists managing NSF programs ought to determine the distribution of funding by scientific field based on their deep knowledge of research merits and needs in each field, and how these needs are best balanced across NSF's research portfolio.

In recent years, a substantial portion of NSF (and other agency) funding has come through continuing appropriations. Continuing appropriations acts—often known as continuing resolutions or CRs—that provide short-term funding until appropriations decisions are finalized can lead to uncertainty for agencies. On one hand, CRs allow for ongoing appropriations discussions without a funding gap. On the other hand, they may lead to reductions or delays in agency operations, such as hiring staff, granting awards and contracts, and beginning new projects, as CRs typically prohibit new activities not funded in the previous fiscal year. Since FY1997, CRs have been enacted on average almost six times per

---

[23] See remarks and presentations materials by Dr. James Ulvestad, Acting Assistant Director of the Mathematical and Physical Sciences (MPS) Directorate, at the NSF MPS Advisory Committee Meeting (MPSAC), June 15-16, 2017, Arlington, VA, available at https://www.nsf.gov/events/event_summ.jsp?cntn_id=191705&org=MPS.

year and provided an average of almost five months of funding annually.[24] The Continuing Appropriations Act, 2018 (P.L. 115-56, Division D), provides funding for NSF through December 8, 2017, at the FY2017 level, subject to a 0.6791% across-the-board decrease.

Further, when funding for an agency and its programs remains at prior year levels, overall purchasing power of appropriated monies is effectively reduced due to the impacts of annual inflation, which has been targeted at 2% by the Federal Reserve in recent years.[25] Both the House and Senate committee reports direct "not less than the fiscal year 2017 enacted level" for multiple programs. As shown in Figure 1 and Figure 2, though current dollar appropriations for NSF have generally increased since FY2010, inflation-adjusted appropriations have remained flat on average. Some analysts argue that even small sustained losses in federal science funding may lead to long-term negative impacts to scientific research and innovation. This may be particularly true for basic research, which often has more uncertainty and longer timelines for generating returns on investment than applied research. Others argue that appropriations levels have remained strong for NSF given the budget constraints of recent years and that funding from other public and private sources should be sought to support scientific research broadly.

---

[24] See CRS Report RL34700, *Interim Continuing Resolutions (CRs): Potential Impacts on Agency Operations*, by Clinton T. Brass; and CRS Report R42647, *Continuing Resolutions: Overview of Components and Recent Practices*, by James V. Saturno and Jessica Tollestrup.

[25] See CRS In Focus IF10477, *Introduction to U.S. Economy: Inflation*, by Jeffrey M. Stupak.

In: Science Policies and Programs
Editor: Johnnie Rodgers
ISBN: 978-1-53614-107-8

Chapter 4

THE PRESIDENT'S FY2018 BUDGET REQUEST FOR THE NATIONAL SCIENCE FOUNDATION*

Laurie A. Harris

The Trump Administration released the FY2018 Budget Request to Congress for the National Science Foundation (NSF) on May 23, 2017, proposing significant funding reductions across the agency's major research, education, and construction accounts. Overall, the request includes $6.653 billion for NSF, $819 million (11%) below the FY2017 enacted amount of $7.472 billion (P.L. 115-31). If funded at the requested level, NSF appropriations would be the lowest since FY2002 in inflation-adjusted (constant) dollars (Figure 1). Ultimately, Congress will determine FY2018 appropriations levels and may incorporate elements of the President's request upon consideration of where, and to what extent, congressional priorities align with those of the Administration.

---

* This is an edited, reformatted, and augmented version of a Congressional Research Service Insight report, IN10718, dated June 14, 2017.

RESEARCH AND EDUCATION

Under the FY2018 request, NSF's main research account, Research and Related Activities (RRA), would receive $5.362 billion, a decrease of $672 million (11.1%) from the FY2017 enacted amount (Table 1). Compared to FY2016 actual amounts, the request would reduce funding for each RRA subaccount except the U.S. Arctic Research Commission (no change). (FY2017 enacted funding was not specified for subaccounts, and the FY2018 request compares subaccount totals to FY2016 actual amounts because FY2017 estimated amounts were not available.) Integrative Activities would receive the largest percentage reduction (26%, $110.8 million decrease), while the remaining subaccounts would receive reductions of between 7.1% and 10.6%.

NSF's main education account, Education and Human Resources (EHR), would receive $760.6 million in FY2018, the largest percentage reduction (13.6%, $119.4 million decrease) requested among NSF's major accounts compared to FY2017 enacted levels. Among EHR's four divisions, compared to FY2016 actual levels, the Division of Graduate Education (DGE) would receive the largest reduction (20.5%, $56.9 million decrease). This includes requested decreases in DGE funding for the CyberCorps: Scholarship for Service program (20%, $10 million decrease) and the Graduate Research Fellowship Program (25.9%, $43 million decrease). The other divisions for undergraduate education, human resourc development, and research on learning would receive reductions of 12%, 9.4%, and 11%, respectively.

CONSTRUCTION AND OPERATIONS

The Major Research Equipment and Facilities Construction (MREFC) account supports major research infrastructure acquisition and construction. The FY2018 budget request would provide $182.8 million for MREFC, a 12.5% ($26.2 million) decrease from the FY2017 enacted amount. Proposed funding includes $105 million to support two new Regional Class Research Vessels (RCRVs), requested prior to congressional direction for $121.9 million to support three RCRVs (Division B Explanatory Statement accompanying P.L. 115-31). The budget request notes that the direction from Congress for three RCRVs will impact current and future funding requirements at unspecified amounts.

FY2018 requested funding levels for the other NSF major appropriations accounts are similar to the FY2017 enacted amounts. The request includes $328.5 million for Agency Operations and Award Management (0.5%, $1.5 million decrease), $4.4 million for the National Science Board (no change), and $15.0 million for the Office of the Inspector General (1.3%, $200,000 decrease).

POTENTIAL POLICY IMPACTS

In March 2017, the House Committee on Science, Space, and Technology, Subcommittee on Research and Technology held hearings to examine NSF policies and programs, focusing on oversight on March 9, and "future opportunities and challenges for science" on March 21. On June 7, 2017, the House Committee on Appropriations, Subcommittee on Commerce, Justice, Science, and Related Agencies held a budget hearing examining proposed NSF FY2018 funding.

According to NSF's estimates, overall funding rates for research grant awards under FY2018 requested funding would decrease from 21% to 19%, compared to FY2016, resulting in 800 fewer grants awarded. Most of NSF's broadening participation and diversity programs would receive reduced funding. However, some programs would receive slight funding increases in FY2018, such as the Inclusion across the Nation of Communities of Learners of Underrepresented Discoverers in Engineering and Science (NSF INCLUDES) program (6.5%, a $1 million increase).

Two programs codified in the 2016 American Innovation and Competitiveness Act (P.L. 114-329) would receive reductions compared to

FY2017 directed funding: the Established Program to Stimulate Competitive Research (EPSCoR) and Innovation Corps (I-Corps). The request includes $100 million for EPSCoR (37.5%, $60 million decrease) and $26.1 million for I-Corps (12.1%, $3.6 million decrease).

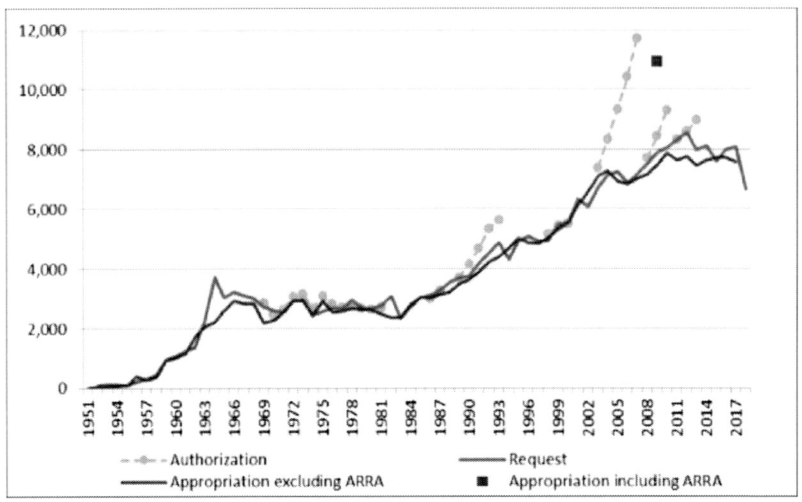

Source: Authorization funding data from selected FY1951 to FY2013 authorization acts. Request and appropriations funding data from the NSF Budget Internet Information System, "NSF Requests and Appropriations History," NSF.gov, June 13, 2016. To calculate constant dollars, CRS used the Gross Domestic Product (Chained) Price Index found in Office of Management and Budget, Historical Tables, "Table 10.1."

Notes: Appropriations funding data are as provided in annual appropriations acts plus adjustments including rescissions, funding transfers across accounts, supplemental appropriations (with American Recovery and Reinvestment Act, P.L. 111-5, FY2009 funding noted separately).

Figure 1. Constant Dollar NSF Authorizations, Budget Requests, and Appropriations: FY1951–FY2018 (in millions of constant dollars (FY2018 estimated)).

Adjusting for inflation, the request would reduce NSF's funding to levels last appropriated prior to the enactment of the America Creating Opportunities to Meaningfully Promote Excellence in Technology, Education, and Science (COMPETES) Act (P.L. 110-69). The COMPETES Act was enacted in 2007 in response to concerns about U.S. competitiveness in scientific and technological innovation, and included authorizations of increased appropriations for NSF. Congress may consider the long-term impacts to research, education and workforce development, and innovation under FY2018 requested funding levels weighed against concerns about the federal budget deficit and debt, and other budget priorities.

Table 1. NSF Funding by Major Account (in millions of dollars)

			Change (FY2018 Request from FY2017 Enacted)		
Account	FY2016 Actual	FY2017 Enacted	FY2018 Request	Dollars	Percentage
Research and Related Activities (RRA)	$5,998.1	$6,033.6	$5,361.7	($672.0)	(11.1%)
Education and Human Resources (EHR)	884.1	880.0	760.6	(119.4)	(13.6)
Major Research Equipment and Facilities Construction (MREFC)	241.5	209.0	182.8	(26.2)	(12.5)
Agency Operations and Award Management (AOAM)	351.1	330.0	328.5	(1.5)	(0.5)
National Science Board (NSB)	4.3	4.4	4.4	0.0	0.0
Office of the Inspector General (OIG)	14.8	15.2	15.0	(0.2)	(1.3)
NSF, Total	$7,493.9	$7,472.2	$6,652.9	($819.3)	(11.0%)

Source: CRS analysis of data from the NSF *FY2018 Budget Request to Congress*, and P.L. 115- 31.

Notes: Parentheses denote negative numbers. Totals may not add due to rounding. Appropriation dollars not adjusted for inflation.

In: Science Policies and Programs
Editor: Johnnie Rodgers

ISBN: 978-1-53614-107-8

Chapter 5

OFFICE OF SCIENCE AND TECHNOLOGY POLICY (OSTP): HISTORY AND OVERVIEW*

John F. Sargent Jr. and Dana A. Shea

ABSTRACT

Congress established the Office of Science and Technology Policy (OSTP) through the National Science and Technology Policy, Organization, and Priorities Act of 1976 (P.L. 94-282). The act states, "The primary function of the OSTP Director is to provide, within the Executive Office of the President [EOP], advice on the scientific, engineering, and technological aspects of issues that require attention at the highest level of Government." Further, "The Office shall serve as a source of scientific and technological analysis and judgment for the President with respect to major policies, plans, and programs of the Federal Government."

The President nominates the OSTP Director, who is subject to confirmation by the Senate. In many Administrations, the President has

* This is an edited, reformatted, and augmented version of a Congressional Research Service report, R43935, dated August 17, 2017.

concurrently appointed the OSTP Director to the position of Assistant to the President for Science and Technology (APST), a position which allows for the provision of confidential advice to the President on matters of science and technology. While Congress can require the OSTP Director to testify, the APST may decline requests to testify on the basis of separation of powers or executive privilege. The APST manages the National Science and Technology Council (NSTC), an interagency body established by Executive Order 12881 that coordinates science and technology (S&T) policy across the federal government. The APST also co-chairs the President's Council of Advisors on Science and Technology (PCAST), a council of external advisors established by Executive Order 13539 that provides advice to the President. Executive Order 13708 continued PCAST through September 30, 2017. As of August 2017, President Trump had not named a director of OSTP or an APST.

Several recurrent OSTP issues face Congress: the need for science advice within the EOP; the title, rank, and responsibilities of the OSTP Director; the policy areas for OSTP focus; the funding and staffing for OSTP; the roles and functions of OSTP and NSTC in setting federal science and technology policy; and the status and influence of PCAST. Some in the S&T community support raising the OSTP Director to Cabinet rank, contending that this would imbue the position with greater influence within the EOP. Others have proposed that the OSTP Director play a greater role in federal agency coordination, priority setting, and budget allocation. Both the Administration and Congress have identified areas of policy focus for OSTP staff, raising questions of prioritization and oversight. Some experts say NSTC has insufficient authority over federal agencies engaged in science and technology activities and that PCAST has insufficient influence on S&T policy; they question the overall coordination of federal science and technology activities. Finally, some in the scientific community support increasing the authority of the OSTP Director in the budget process to bring greater science and technology expertise to federal investment decision making.

Historically, advice to the President was provided through advisors and boards without statutory authorities. Congress moved in 1976 to codify a formal mechanism for presidential science advice. The National Science and Technology Policy, Organization, and Priorities Act of 1976 (P.L. 94-282) established the Office of Science and Technology Policy (OSTP), including the position of its Director, within the Executive Office of the President (EOP) to provide scientific and technological analysis and advice to the President. This act codified and institutionalized a presidential science advice function that previously existed at each President's discretion.

This chapter provides an overview of the history of science and technology (S&T) advice to the President and discusses selected recurrent issues for Congress regarding OSTP's Director, OSTP management and operations, the President's Council of Advisors on Science and Technology (PCAST), and the National Science and Technology Council (NSTC).

HISTORY OF SCIENCE AND TECHNOLOGY ADVICE TO THE PRESIDENT

Science and technology policy issues tend to reach the presidential level if they involve multiple agencies; have substantial budgetary, economic, national security, or foreign policy dimensions; are highly controversial (especially when science and technology intersect with values, ethics, and morality); or are highly visible to the public. When these matters reach the Oval Office, Presidents generally seek information and advice from trusted sources as to the options available and their implications.

Throughout U.S. history, Presidents have obtained S&T advice from federal scientists and engineers and informal personal contacts.[1] Starting in the early 1930s, Presidents attempted to expand their sources of S&T advice through advisory boards and committees. Lacking a statutory foundation, these boards and committees tended to lack permanency, as subsequent Presidents often disbanded them. When again faced with the need for S&T advice, Presidents would form new advisory boards or committees, sometimes reconstituted from previously disbanded ones.

---

[1] For a history of OSTP, see Genevieve J. Knezo, "Science and Technology," Chapter 6 in Harold C. Relyea (ed.), *The Executive Office of the President: A Historical, Biographical, and Bibliographical Guide* (Westport, Connecticut: Greenwood Press, 1997).

In the years leading up to World War II, the importance of research and development (R&D) to the nation's economic and military strength became increasingly evident. As a result, President Franklin D. Roosevelt established the Office of Scientific Research and Development (OSRD) in 1941.[2] The federal R&D enterprise is widely credited with contributing substantially to the Allied victory in World War II, as well as to the development of subsequent U.S. industrial strength.[3] In November 1944, President Roosevelt wrote a letter to OSRD Director Vannevar Bush[4] seeking recommendations on how research and the research infrastructure established to support America's war effort could be "profitably employed in times of peace."[5] Bush's response, Science: The Endless Frontier,[6] laid out a framework that asserted the essential role of scientific progress in meeting the nation's economic, national security, and social needs. Experts widely view the Bush report as foundational to today's U.S. science and technology policy.

Subsequent Presidents used a variety of mechanisms to obtain S&T advice within the EOP, to enhance interagency coordination, and to receive counsel from outside advisors. The primary provision of advice to the President on science and technology issues continued through advisors and assistants to the President who continued to perform this function without statutory authorities. Organizations within the EOP included the Office of

---

[2] President Roosevelt established OSRD within the Office for Emergency Management of the Executive Office of the President. Executive Order 8807, "Establishing the Office of Scientific Research and Development," June 28, 1941, http://www.presidency.ucsb.edu/ws/?pid=16137.

[3] See, for example, William A. Blanpied, "Science Policy in the Early New Deal and Its Impacts in the 1940s," *Federal History online*, January 2009, pp. 9-24, and John Brooks Slaughter, "National Science Foundation," in *Encyclopedia of Education Economics and Finance* (SAGE Publications, 2014), p. 477.

[4] OSRD Director Bush reported directly to President Roosevelt.

[5] Letter from President Franklin D. Roosevelt to Vannevar Bush, Director, Office of Scientific Research and Development, November 17, 1944, http://www.nsf.gov/od/lpa/nsf50/vbush1945.htm#letter.

[6] Vannevar Bush, *Science The Endless Frontier: A Report to the President by Vannevar Bush, Director of the Office of Scientific Research and Development*, Office of Scientific Research and Development, Executive Office of the President, Washington, DC, July 5, 1945, http://www.nsf.gov/od/lpa/nsf50/vbush1945.htm#ch1.

the Special Assistant to the President for Science and Technology (Eisenhower) and the Office of Science and Technology (OST; Kennedy, Johnson). Organizations focused on interagency coordination included the President's Scientific Research Board (Truman), the Federal Council for Science and Technology (FCST; Eisenhower, Kennedy, Johnson, Nixon), and the Federal Coordinating Council for Science, Engineering, and Technology (FCCSET; Ford, Carter, Reagan, George H. W. Bush). External advisory committees included the Science Advisory Committee (Truman, Eisenhower), and the President's Science Advisory Committee (PSAC; Eisenhower, Kennedy, Johnson, Nixon).

In 1973, President Nixon abolished the Office of Science and Technology. The National Science Foundation (NSF) assumed its civilian functions and the National Security Council (NSC) its security functions.[7] In addition, President Nixon opted not to appoint new members to PSAC after accepting the pro forma resignation of its members.[8] With this backdrop, President Ford chose to establish OSTP through legislation, rather than executive order.[9] The National Science and Technology Policy, Organization, and Priorities Act of 1976 (P.L. 94-282) established OSTP and the position of OSTP Director. President Ford signed it into law on May 11, 1976.

The creation of OSTP provided a new structure for the provision of science and technology policy advice to the President, but did not end Presidents' authority to appoint advisors in parallel. The OSTP director is a statutory position; the authority to appoint others to assist the President exists solely with the President. Thus, a President may opt to appoint the OSTP director to also serve as an assistant to the President, may concurrently appoint another individual to serve as Assistant to the

---

[7] David Z. Beckler, "The Precarious Life of Science in the White House," *Daedalus*, vol. 103, no. 3 (Summer 1974), p. 115, http://www.jstor.org/stable/20024223.

[8] Ibid.

[9] Jeffrey K. Stine, *A History of Science Policy in the United States, 1940-1985*, Report for the House Committee on Science and Technology Task Force on Science Policy, 99th Cong., 2nd sess., Committee Print (Washington, DC: GPO, 1986), http://ia341018.us.archive.org/2/items/historyofscience00unit/historyofscience00unit.pdf. See also Roger Pielke, and Roberta A. Klein (Editors), *Presidential Science Advisors Perspectives and Reflections on Science, Policy and Politics*, (New York: Springer, 2010).

President for Science and Technology (APST), or may appoint no one to serve as APST. This also raised new and continuing questions with respect to coordination of advice.

Appendix A provides a historical compilation of presidential S&T policy advisors with their titles, EOP S&T agencies/offices, interagency coordination organizations, and advisory committees. As illustrated in Table A-1, the Presidents subsequent to President Ford continued to adapt OSTP and related organizations to suit their needs.

OVERVIEW OF OSTP, NSTC, AND PCAST

The White House contains several science and technology policy entities, including OSTP, the National Science and Technology Council (NSTC), and the President's Council of Advisors on Science and Technology (PCAST). This section describes the structure, roles and responsibilities, current structure, and budget of each entity. The role and influence of OSTP, NSTC, PCAST, and their predecessor organizations have varied among Administrations, depending on the President, the individual serving as OSTP Director, and the rapport between them.[10]

OFFICE OF SCIENCE AND TECHNOLOGY POLICY

Overview

Congress established the Office of Science and Technology Policy as an office within the EOP to, among other things, "serve as a source of scientific and technological analysis and judgment for the President with

---

[10] For a discussion of the varying influence of science advisors, listen to National Public Radio, *The Evolving Role of the Presidential Science Advisor*, Talk of the Nation, November 16, 2007, http://www.npr.org/templates/story/ story.php?storyId=16343713.

respect to major policies, plans, and programs of the Federal Government."[11] OSTP defines its mission as having three components:

> Provide the President and his senior staff with accurate, relevant, and timely scientific and technical advice on all matters of consequence.
>
> Ensure that the policies of the Executive Branch are informed by sound science.
>
> Ensure that the scientific and technical work of the Executive Branch is properly coordinated so as to provide the greatest benefit to society.[12]

To accomplish this mission, OSTP has established the following strategic goals and objectives:

> Ensure that federal investments in science and technology are making the greatest possible contribution to economic prosperity, public health, environmental quality, and national security.
>
> Energize and nurture the processes by which government programs in science and technology are resourced, evaluated, and coordinated.
>
> Sustain the core professional and scientific relationships with government officials, academics, and industry representatives that are required to understand the depth and breadth of the Nation's scientific and technical enterprise, evaluate scientific advances, and identify potential policy proposals.
>
> Generate a core workforce of world-class expertise capable of providing policy-relevant advice, analysis, and judgment for the President and his senior staff regarding the scientific and technical aspects of the major policies, plans, and programs of the Federal government.[13]

The OSTP also has several roles not articulated in these formal statements. These include serving as a sounding board and conduit of information for agency executives seeking to understand, clarify, and shape science and technology-related policy objectives and priorities; helping agencies coordinate and integrate their S&T strategies and

---

[11] P.L. 94-282.

[12] OSTP, "About OSTP," http://www.whitehouse.gov/administration/eop/ostp/about.

[13] Ibid.

activities; and helping resolve interagency conflicts over areas of S&T responsibility and leadership.

OSTP Structure/Roles of the OSTP Director, APST, and Associate Directors

Past Presidents appointed Assistants to the President for Science and Technology (or their equivalents) to coordinate presidential advice. Congress codified a specific science and technology advisory function when it created OSTP. P.L. 94-282 establishes the position of OSTP Director, whose primary function is

> to provide, within the Executive Office of the President, advice on the scientific, engineering, and technological aspects of issues that require attention at the highest level of Government.

In addition, the statute, as amended,[14] directs the OSTP Director to

> advise the President of scientific and technological considerations involved in areas of national concern including, but not limited to, the economy, national security, homeland security, health, foreign relations, the environment, and the technological recovery and use of resources;
>
> evaluate the scale, quality, and effectiveness of the federal effort in science and technology and advise on appropriate actions;
>
> advise the President on scientific and technological considerations with regard to federal budgets, assist the Office of Management and Budget (OMB) with an annual review and analysis of funding proposed for research and development in budgets of all federal agencies, and aid [OMB] and the agencies throughout the budget development process; and
>
> assist the President in providing general leadership and coordination of the research and development programs of the Federal Government.

By statute, the President appoints the OSTP Director, who is sometimes referred to colloquially as the President's science advisor.[15] The

[14] Section 1712(1) of P.L. 107-296 inserted "homeland security" after "national security" in the list of areas of national concern.

OSTP Director is subject to Senate confirmation and receives compensation at the rate provided for level II of the Executive Schedule. The OSTP Director has never been a member of the President's Cabinet or a Cabinet-level official. The statute does not require, nor may Congress compel, that the President appoint the OSTP director to serve as an assistant to the President (or, more specifically, as APST).

In addition to establishing the position of OSTP Director, P.L. 94-282 authorizes the President to appoint not more than four OSTP Associate Directors, subject to Senate confirmation, who are compensated at a rate not to exceed that provided for level III of the Executive Schedule. President Obama established four OSTP Associate Director positions with discrete areas of responsibility: science; technology and innovation; national security and international affairs; and environment and energy. See Figure 1.[16] The number of Associate Director positions has varied under different Presidents. For example, under President George W. Bush there were two OSTP Associate Directors—one focused on science and the other on technology—each with a Deputy Director.[17] During the Clinton Administration, four Associate Directors focused on science; technology; environment; and national security and international affairs. Under President Obama there were four Associated Directors focused on science; technology and innovation; environment and energy; and national security and international affairs. The section "Number and Policy Foci of OSTP

---

[15] Although there is no statutory EOP title or position of "Science Advisor" or "Presidential Science Advisor," this term is often used to describe the individual serving as the primary advisor to the President on science and technology issues. Executive Order 13539 ("President's Council of Advisors on Science and Technology," April 21, 2010) identifies the Assistant to the President for Science and Technology (APST) as the "Science Advisor" and states that the APST shall serve as a co-chair of PCAST; the position of PCAST co-chair is currently vacant.

[16] As of July 2017, OSTP was unable to provide CRS with an update to Figure 1 based on changes implemented or planned by the Trump Administration. OSTP informed CRS that "OSTP's structure is evolving and streamlining to cover ongoing requirements and new priorities for this Administration. Currently, OSTP's work falls into three main work streams: Science, Technology, and National Security. The work of the National Science and Technology Council continues and the President's Council of Advisors on Science and Technology Policy will be re-chartered. Once the new OSTP Director is on board, OSTP will work to implement the Director's vision for how best to structure OSTP." Email communication from OSTP to CRS, July 17, 2017.

[17] CRS discussions with Stanley Sokul, Chief of Staff, Bush Administration OSTP, August 14, 2008.

Associate Directors" provides a more detailed discussion of the role of OSTP Associate Directors.

Ted Wackler currectly serves as OSTP's Acting Director and also leads the NSTC.

Presidential Appointment Status and Congress

The formal positions held by a President's science advisor may affect his or her degree of access to the President and other EOP decisionmakers. Although Presidents have differed in their management of EOP staff, Cabinet members and assistants to the President generally have greater access to the President than other White House staff.[18] The OSTP Director is not a Cabinet-level official.

Some members of the S&T policy community question the degree of presidential access available to the OSTP Director. Some Presidents have appointed their science advisors not only to the Senate-confirmed position of OSTP Director, but also as Assistant to the President for Science and Technology (APST). The APST position does not require Senate confirmation and may confer additional status and access to the President. Presidents Obama and Clinton appointed their OSTP Directors as APST; President George W. Bush did not appoint an APST.

The relationship between Congress and the individual serving as the President's science advisor depends on whether the individual serves as OSTP Director, APST, or both. Congress can require the OSTP Director to testify before Congress. In contrast, APSTs may assert the right not to testify before Congress in accordance with the principles of separation of powers or executive privilege. Congress's authority to require testimony from an individual who holds both the Director of OSTP and APST title may be ambiguous depending on the capacity in which the individual would testify and the subject matter of the testimony.[19]

---

[18] Information on the President's Cabinet is available at http://www.whitehouse.gov/government/cabinet.html.

[19] For a fuller discussion of this issue, see CRS Report RL31351, *Presidential Advisers' Testimony Before Congressional Committees: An Overview*, by Todd Garvey and Henry B. Hogue.

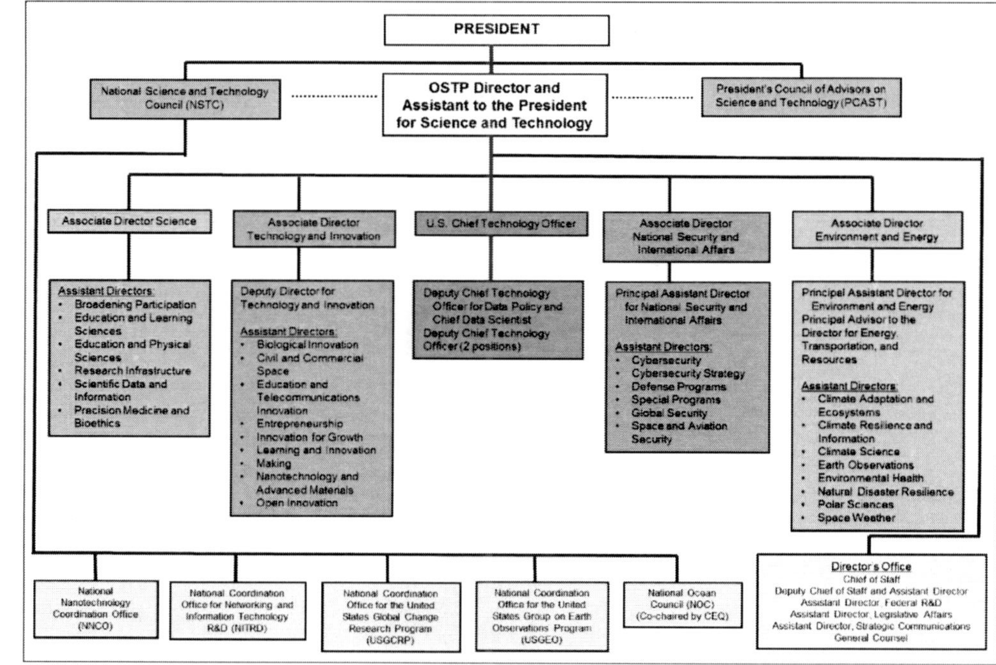

Source: Office of Science and Technology Policy, Executive Office of the President, email communication, April 5, 2016.

Note: As of July 2017, OSTP was unable to provide CRS with an update to Figure 1 based on changes implemented or planned by the Trump Administration. OSTP informed CRS that "OSTP's structure is evolving and streamlining to cover ongoing requirements and new priorities for this Administration." Email communication from OSTP to CRS, July 17, 2017.

Figure 1. Selected White House Science and Technology Policy Organizations as Organized Under President Obama.

Roles and Responsibilities

The OSTP Director advises the President on policy formulation; presidential appointments; S&T-related budget issues, including budgets for R&D; the policy significance of scientific and technical developments; and science, technology, engineering, and mathematics (STEM) education. OSTP Directors historically have also served as communication conduits between the EOP and the federal and non-federal S&T community. Some OSTP Directors have emphasized communicating the views of the S&T community to the EOP, while others have focused on communicating the views of the EOP to the S&T community.

The APST manages the National Science and Technology Council (NSTC), established by Executive Order 12881,[1] which is charged with coordinating S&T policy across the federal government, establishing national goals for federal S&T investments, and preparing coordinated R&D strategies. As NSTC manager, the APST can provide federal agency coordination, information, and guidance when special events occur, such as national emergencies, disasters, or S&T-related international negotiations.

In addition, the APST co-chairs the President's Council of Advisors on Science and Technology (PCAST), established in its current form under President Obama by Executive Order 13539.[2] As co-chair of PCAST, the APST can seek to ascertain the consensus of the S&T community on issues of interest to the Administration.

The OSTP Director performs special roles with respect to national security and emergency preparedness $_2$(NS/EP) communications policies, programs, and capabilities. Under Executive Order 13618,[3] the OSTP Director advises the President on the prioritization of radio spectrum and wired communications that support NS/EP communications functions, and provides selected evaluation of appropriate information related to the test,

---

[1] Executive Order 12881, "Establishment of the National Science and Technology Council," November 23, 1993, http://www.archives.gov/federal-register/executive-orders/pdf/12881.pdf.

[2] Executive Order 13539, "President's Council of Advisors on Science and Technology," April 21, 2010, http://www.gpo.gov/fdsys/pkg/FR-2010-04-27/pdf/2010-9796.pdf.

[3] Executive Order 13618, "Assignment of National Security and Emergency Preparedness Communications Functions," July 11, 2012, http://www.gpo.gov/fdsys/pkg/FR-2012-07-11/pdf/2012-17022.pdf.

exercise, evaluation, and readiness of the capabilities of existing and planned NS/EP communications. In addition, the OSTP Director issues priorities annually for NS/EP Executive Committee analyses, studies, research, and development regarding NS/EP communications.[4]

Relationship with the Office of Management and Budget

The OSTP Director does not have direct authority over federal agencies or the Office of Management and Budget (OMB). OSTP's participation with OMB in the budget process involves four steps: (1) overall priority setting by OSTP and OMB, (2) agency preparation of budget proposals to OMB, (3) agency negotiations with OMB, and (4) final budget decisions by the President and the OMB Director.

1) Priority setting. A key activity in the first step is OSTP's request to federal agencies for their recommendations on R&D priorities. In addition, interagency working groups meet to determine individual agency responsibilities for specific activities when multiple agencies share responsibility for broad issue areas. The OSTP and OMB use this information in their development of a joint memorandum that articulates the Administration's R&D priorities and R&D investment criteria.[5] Agencies are encouraged to use this memorandum as an aid in the second step, preparation of their budgets.
2) Agency budget preparation. In the second step, OSTP continually interacts with agencies as they develop their budgets, providing advice and working with them on their priorities. In general, OSTP provides more guidance to agencies with large R&D budgets and to programs that cross agency boundaries. Federal agencies submit their completed budget proposals to OMB. The OSTP does not review proposed agency budgets before they are sent to OMB.

---

[4] Email communication from OSTP General Counsel Rachael Leonard to CRS, December 11, 2013.

[5] On July 18, 2014, OMB and OSTP issued a joint memorandum on science and technology priorities for FY2016 (http://www.whitehouse.gov/sites/default/files/microsites/ostp/m-14-11.pdf).

3) Agency negotiations with OMB. In the third step, OSTP works with OMB to review proposed agency budgets to ensure they reflect Administration plans and priorities. The OSTP also participates in OMB budget examiner presentations to the OMB Director and provides advice on priorities at that time. In addition, OSTP provides direct feedback to agencies as they negotiate with OMB over funding levels and the programs on which that funding is to be spent.
4) Final budget decisions. OSTP's primary role in the fourth step of the budget process is to advise on the quality of the agency budget proposals and their alignment with the President's established priorities. The President, the OMB Director, and the Cabinet, however, make the ultimate choices.

Budget and Staffing

OSTP's budget and staffing affect the degree to which OSTP can provide advice to the President and respond to congressional direction and mandates. Figure 2 shows OSTP's budget from FY1990 to FY2018 (request), and Figure 3 shows OSTP's staffing level from FY1990 to FY2018 (request). The President's request for OSTP for FY2017 is $5.544 million, a decrease of $11,000 (0.2%) below the FY2017 level provided by P.L. 115-31, and 33 FTE, equal to the FY2017 level.

In FY2012, Congress reduced funding for OSTP substantially; contemporaneously, the Administration transferred responsibility for funding PCAST to the Department of Energy. Funding for support of PCAST, provided by the Department of Energy (DOE) beginning in FY2012, is included in Figure 2. DOE PCAST funding has ranged from $217,000 in FY2013 to $751,000 in FY2015. Funding in FY2016 was $541,000. In FY2017, funding is estimated to be $230,000. Funding supports PCAST salaries and benefits, committee member travel, meeting planning support, and related expenses, and is provided through the DOE Science account. In its FY2018 budget, there is no specific request for PCAST funding:

> The PCAST advisory committee has dissolved and [the DOE Office of Science] is not aware of any plans to reform this committee in FY 2018.[6]

The OSTP is also supported by a federally funded research and development center (FFRDC), the Science and Technology Policy Institute (STPI; see box below), which is staffed and funded through the National Science Foundation appropriation. The President is requesting $4.0 million for STPI for FY2018; funding for FY2017 in the NSF current plan is $4.74 million.[7]

As illustrated in Figure 2 and Figure 3, OSTP funding and staffing levels have varied considerably over time. In constant dollars, OSTP funding was at its highest at the end of the George H. W. Bush Administration and at its lowest during the Reagan Administration (see Figure B-1, which illustrates OSTP funding since 1977). OSTP's staffing has also fluctuated.

As of July 17, 2017, OSTP had a total of 36 staff members covering OSTP's portfolio of work. This includes three political staff, 14 career staff, one unpaid consultant, 15 detailees, two IPAs, and one fellow.[8] Additionally, OSTP reports that it "has been actively recruiting and hiring new staff members who will be arriving on a rolling basis."[9] Generally, political staff and career staff are funded by OSTP; detailees are funded by their home agencies; fellows are funded by a variety of organizations; and

---

[6] Department of Energy, *FY2018 Department of Energy's Budget Request to Congress, Volume 4, Science*, p. 339.

[7] Email communication between NSF and CRS, August 14, 2017.

[8] Email communication from OSTP to CRS, July 27, 2017. A detail is an officially approved temporary assignment of a civil service employee (informally called a "detailee") to a different position in another federal agency; the employee's official title, series, grade, rate of compensation, and permanent employer do not change. The Office of Personnel Management's Intergovernmental Personnel Act Mobility Program provides for the temporary assignment of personnel (IPAs) between the federal government and state and local governments, colleges and universities, Indian tribal governments, federally funded research and development centers, and other eligible organizations. Fellows are scientists and engineers who come to Washington, DC, to gain experience in public policy and provide science and technical advice to policymakers. Most are recent graduates of doctoral programs, but some are more experienced staff from industry or universities. Fellows generally come for one year, but that time can be extended.

[9] Email communication from OSTP to CRS, July 27, 2017.

IPAs may be funded by OSTP, their home agencies/organizations, or a combination of the two.[10]

OSTP funds its political and career staff, and includes relevant information in its annual budget requests to Congress. Additionally, OSTP has relied on detailees, fellows, and IPAs to support its activities for at least the last three presidential administrations. Detailees, fellows, and IPAs may be funded by OSTP, their home agencies/organizations, or a combination of the two. During the Obama Administration, OSTP began with approximately 30 and ended with approximately 70 detailees, IPAs, and fellows. During the G.W. Bush Administration, OSTP had approximately 30- 40 detailees per year. Toward the end of the Clinton Administration, OSTP had approximately 60 detailees and fellows.[11]

Science and Technology Policy Institute

The Science and Technology Policy Institute (STPI) is a federally funded research and development center (FFRDC) that provides analytical support to the Office of Science and Technology Policy, the National Science Foundation (NSF), and the National Science Board. Congress created STPI through the National Defense Authorization Act for Fiscal Year 1991 (P.L. 101-510). This law established the Critical Technologies Institute (CTI), an FFRDC under the sponsorship of OSTP but supported by appropriations provided to the Department of Defense (DOD). The RAND Corporation initially managed CTI. In 1998, Congress enacted the National Science Foundation Authorization Act of 1998 (P.L. 105-207), which changed CTI's name to the Science and Technology Policy Institute, changed primary sponsorship to the National Science Foundation, and amended the institute's duties.

In 2003, the Institute for Defense Analysis (IDA) was selected to

[10] Office of Science and Technology Policy, personal communication, March 23, 2016. In an earlier email (January 24, 2012) to CRS, OSTP General Counsel Rachael Leonard asserted that OSTP may reimburse agencies for all or part of the personnel costs, but is not required to do so under the terms of 3 U.S.C. 112, the provisions of which apply only to the White House Office, the Executive Residence at the White House, the Office of the Vice President, the Domestic Policy Staff, and the Office of Administration.

[11] Email communication from OSTP to CRS, July 27, 2017.

manage STPI. NSF appropriations provides funding for STPI, including $4.7 million in FY2017. The STPI has approximately 40 full-time employees.a The STPI may also contract for expertise as required for a particular project.b In addition, STPI has access to the expertise of IDA's approximately 800 other employees.

The duties of STPI include:

1) The assembly of timely and authoritative information regarding significant developments and trends in science and technology research and development in the United States and abroad.
2) Analysis and interpretation of the information referred to in paragraph (1) with particular attention to the scope and content of the federal science and technology research and development portfolio as it affects interagency and national issues.
3) Initiation of studies and analysis of alternatives available for ensuring the long-term strength of the United States in the development and application of science and technology, including appropriate roles for the federal government, state governments, private industry, and institutions of higher education in the development and application of science and technology.
4) Provision, upon the request of the Director of the Office of Science and Technology Policy, of technical support and assistance
   a) to the committees and panels of the President's Council of Advisors on Science and Technology that provide advice to the Executive Branch on science and technology policy; and
   b) to the interagency committees and panels of the federal government concerned with science and technology. C

In carrying out these duties, the statute directs STPI to consult widely with representatives from private industry, academia, and nonprofit institutions, and to incorporate their views in STPI's work to the maximum extent practicable. In addition, the statute requires STPI to submit an annual report to the President on its activities, in accordance with requirements prescribed by the President.

In addition to its primary customer, OSTP, and its sponsor, NSF, STPI has conducted work for other federal agencies including: the National Institutes of Health; Department of Transportation; DOD; Department of Health and Human Services; National Science Board; Department of Commerce, including the National Institute of Standards and Technology; Department of Homeland Security; and Department of Energy.

a) Full-time employees are defined as those with approximately 80% or more of their work time devoted to STPI work.
b) Email communication from STPI Deputy Director Bill Brykczynski to CRS, February 26, 2016.
c) 42 U.S.C. 6686.

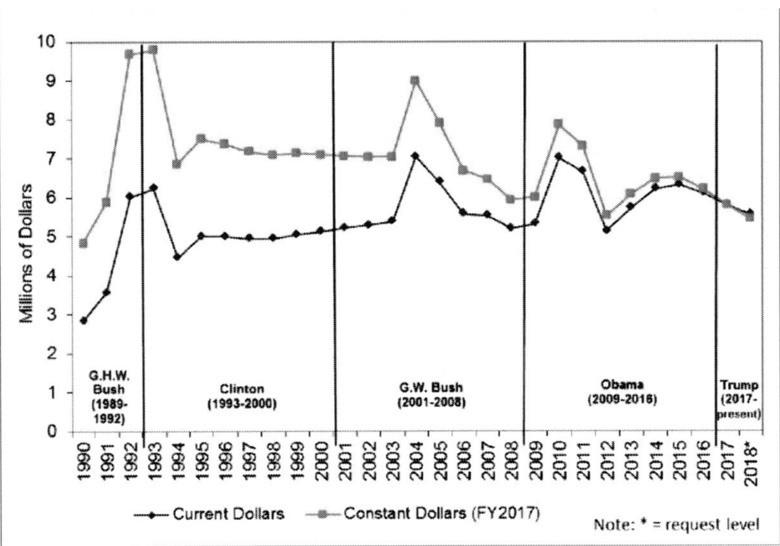

Sources: Congressional Research Service. Data from OMB Public Budget Database, budget requests, and congressional appropriations acts and committee reports, FY1990-FY2018; PCAST funding data from the Department of Energy, emails from DOE to CRS.

Notes: In FY2008, Congress directed NSF to transfer $2.240 million to OSTP for Science and Technology Policy Institute (STPI) (not shown). If the STPI funding were included, FY2008 funding for OSTP would be $7.424 million in current dollars. The data above includes funding for PCAST provided by DOE starting in FY2012.

Figure 2. OSTP Funding, FY1990-FY2017 (actual), FY2018 (request).

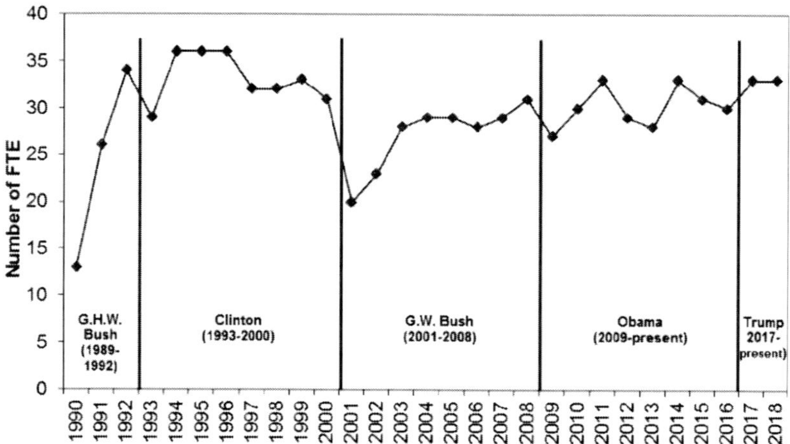

Sources: Congressional Research Service. Data from U.S. Office of Management and Budget, *Budget of the United States Government*, Appendix, FY1992-FY2018. (Note that actual staffing numbers are provided two years later. For example, to determine actual staffing in FY2007, one must review the FY2009 budget request.) The OMB did not provide this data for FY200 1. CRS has estimated the number of FTEs for FY200 1 based on information provided by OSTP. FY2017 and FY2018 figures are based on requests.

Notes: Data reported are in full-time equivalents (FTE, the amount of effort from one full-time employee over one year) and may not equal number of staff. Data do not include staff or FTEs funded by agencies other than OSTP, such as detailees, IPAs, and fellows. Historical data includes full-time equivalent of holiday and overtime hours.

Figure 3. OSTP Staffing, FY1990-FY2016 Actual, FY2017 (est.), FY2018 (request).

National Science and Technology Council

Overview and Structure

On November 23, 1993, President Clinton established the NSTC by Executive Order 12881.[12] The NSTC is composed of department and agency heads, as well as selected assistants and advisors to the President. Executive Order 12881 specifies that the APST is a member of the NSTC; the order does not include the OSTP Director in the NSTC membership.

[12] Executive Order 12881, "Establishment of the National Science and Technology Council," 58 *Federal Register* 62491-62492, November 23, 1993.

The NSTC aims to coordinate science and technology policy across the federal government. According to the executive order, the NSTC has the following principal functions:

- coordinate the S&T policymaking process;
- ensure S&T policy decisions and programs are consistent with the President's stated goals;
- help integrate the President's S&T policy agenda across the federal government;
- ensure science and technology are considered in development and implementation of federal policies and programs; and
- further international cooperation in science and technology.

In addition to these principal functions, the NSTC assists the OMB Director by recommending R&D budgets that reflect national goals and advising on agency R&D submissions.

The President chairs the NSTC; in the President's absence, the Vice President or the APST serves as chair. In practice, the NSTC rarely meets with the President or Cabinet-level officials present. Rather, OSTP staff and detailees implement NSTC activities in conjunction with federal agency staff. According to OSTP, Ted Wackler is Acting Director of OST and is leading the NSTC.[13]

Under President Obama, the NSTC had five committees: Science; Technology; Environment, Natural Resources, and Sustainability; Homeland and National Security; and Science, Technology, Engineering, and Math (STEM) Education.[14] As shown in Table 1, each NSTC committee has subcommittees, interagency working groups, and/or taskforces focused on specialized topics. The members of these committees and subcommittees are generally not Cabinet officials, but instead lower-ranking staff.

---

[13] Email from OSTP to CRS, August 15, 2017.

[14] OSTP asserts that "The work of the National Science and Technology Council continues and the President's Council of Advisors on Science and Technology Policy will be re-chartered. Once the new OSTP Director is on board, OSTP will work to implement the Director's vision for how best to structure OSTP." Email from OSTP to CRS, July 27, 2017.

Table 1. National Science and Technology Council Committees Under President Obama

COMMITTEE ON ENVIRONMENT, NATURAL RESOURCES, AND SUSTAINABILITY (CENRS) Tamara Dickinson (OSTP), Kathryn Sullivan (NOAA), Thomas Burke (EPA)		
AQRS: Air Quality Research (SC)		SOST: Ocean Science & Technology (SC)*
CSMSC: Critical & Strategic Mineral Supply Chains (SC)	Water-Energy-Food (TF)	SWAQ: Water Availability & Quality (SC)
IARPC: Interagency Arctic Research Policy Committee (IWG)*	SES: Ecological Systems(SC)	T&R: Toxics & Risk (SC)
SDR: Disaster Reduction (SC)	SGCR: Global Change Research (SC)*	USGEO: U.S. Group on Earth Observations (SC)
COMMITTEE ON HOMELAND & NATIONAL SECURITY (CHNS) Steve Fetter(OSTP), Steve Welby (DoD), Reginald Brothers (DHS)		
BDRD: Biological Defense Research & Development (SC)	Astronomical Assets and Data (IWG)	SCORE: Subcommittee on Special Cyber Operations Research and Engineering (SC)
CDRD: Chemical Defense Research and Development (SC)	NDRD: Nuclear Defense Research & Development (SC)	TISTI: Topics in International Science, Technology and Innovation (SC)
STCED: Science and Technology for Countering Explosive Devices (SC)	National Security Laboratory Research, Development, Test and Evaluation Facilities and Infrastructure (SC)	FTAC for Counter Autonomous and Unmanned Vehicles (FTAC)
SOS-CBRNE Standards (SC)	CISR: Critical Infrastructure Security and Resilience (SC)	AUS: Autonomous Unmanned Systems (SC)
	DAMIEN: Detecting & Mitigating the Impact of Earth-Bound Near Earth Objects (IWG)	
COMMITTEE ON SCIENCE (CoS) Francis Collins (NIH), Jo Handelsman (OSTP), France Cordova (NSF)		
IWGN: Neuroscience (IWG)*	PSSC: Physical Science (SC)	LSSC: Life Science (SC)*
Social, Behavioral, and Economic Science (SC)	IWGMI: Medical Imaging (IWG)*	SMDIS: Strengthening the Medicolegal Death Investigation System (FTAC)
COMMITTEE ON STEM EDUCATION (CoSTEM)* Jo Handelsman (OSTP), France Cordova (NSF)		
FC-STEM: Federal Coordination in STEM Education (SC)		
COMMITTEE ON TECHNOLOGY (CoT) Thomas Kalil (OSTP)		
ASTS: Aeronautics Science & Technology (SC)	SAM: Advanced Manufacturing (SC)*	SoS: Standards (SC)
GIG: Global Internet Governance (SC)	DGT: Digital Game Technologies (IWG)	MGI: Material Genome Initiative (SC)
Social and Behavioral Sciences Team (SC)	NITRD: Network and Information Technology R&D (SC)*	NSET: Nanoscale Science Engineering & Technology (SC)*
	Lab to Market (SC)	
TASK FORCE ON ZIKA S&T RESPONSE Jo Handelsman (OSTP)		

Source: OSTP, EOP, email communication, April 5, 2016.

Notes: As of July 2017, OSTP was unable to provide CRS with an update to Table 1 based on changes implemented or planned by the Trump Administration. OSTP informed CRS that "The work of the National Science and Technology Council continues." Email communication from OSTP to CRS, July 17, 2017. A red asterisk indicates a congressionally mandated group or subgroup.

Underlining indicates a group overseeing an initiative that is supported by a National Coordination Office. SC = subcommittee; IWG = interagency working group; TF = task force.

In some cases, Congress has charged the NSTC with specific statutory responsibilities. Congress mandated the NSTC to coordinate federal activities on ocean acidification[15] and to develop an implementation plan for a coordinated national research program on the role of the oceans in human health and report annually on these activities.[16] Congress also directed the NSTC to oversee the planning, management, and coordination of the National Nanotechnology Program and report annually on these activities.[17] In addition, Congress directed the OSTP Director to establish an NSTC committee responsible for coordinating federal programs and activities in support of STEM education,[18] to establish a committee responsible for planning and coordinating federal programs and activities in advanced manufacturing research and development,[19] to establish a working group responsible for coordinating federal science agency research and policies related to the dissemination and long-term stewardship of the results of unclassified research,[20] and to use the NSTC to annually identify and prioritize deficiencies in federal research facilities and major instrumentation.[21]

In other cases, the NSTC may be assigned responsibilities to meet non-specific congressional mandates. For example, the America COMPETES Act (P.L. 110-69) directs the establishment of a President's Council on Innovation and Competitiveness. The act states that the council is to include the Secretary or head of a number of federal agencies, OSTP, and OMB. Congress provided the President with the option of establishing a new organization to serve as the Council on Innovation and Competitiveness or to designate an existing council to carry out the requirement. Rather than establish a new, independent council, President

---

[15] P.L. 111-11, "The Omnibus Public Land Management Act of 2009," §12403.

[16] P.L. 108-447, Division B, Title IX, "Oceans and Human Health Act," §902.

[17] P.L. 108-153, §2, "21st Century Nanotechnology Research and Development Act." The act refers to a National Nanotechnology Program, but the broader effort is generally referred to in the executive branch as the National Nanotechnology Initiative or NNI.

[18] P.L. 111-358, "America COMPETES Reauthorization Act of 2010," §101.

[19] P.L. 111-358, "America COMPETES Reauthorization Act of 2010," §102.

[20] P.L. 111-358, "America COMPETES Reauthorization Act of 2010," §103.

[21] P.L. 110-69, "America COMPETES Act," §1007.

George W. Bush assigned this responsibility to the NSTC Committee on Technology.[22]

Budget and Staffing

The NSTC receives no direct appropriations. Instead, the participating agencies provide funding that the NSTC uses to coordinate multi-agency programs. The amount provided varies and has ranged from approximately $12 million to $18 million from FY2010 to FY2015; funding was $17.9 million in FY2014 and $18.1 million in FY2015. This amount excludes infrastructure contributions from OSTP and funding for NSTC activities that are solely within a single agency. NSTC staff are assigned by their agencies. The number of NSTC assignees has varied from 5 in prior years up to 21 in FY2015. More current data will be available upon publication of a FY2016 NSTC report which is in the final stages of clearance.[23]

President's Council of Advisors on Science and Technology

Overview and Structure

President George H. W. Bush created the President's Council of Advisors on Science and Technology (PCAST) in 1990.[24] Presidents Clinton, George W. Bush, and Obama reestablished slightly different versions of PCAST during their Administrations.[25] According to OSTP, President Trump will re-charter PCAST.[26] PCAST was last extended by

---

[22] Memorandum of the President of the United States, "Designation of the Committee on Technology of the National Science and Technology Council to Carry Out Certain Requirements of the America COMPETES Act," 73 *Federal Register* 20523, April 10, 2008.

[23] Email from OSTP to CRS, August 15, 2017.

[24] Executive Order 12700, "President's Council of Advisors on Science and Technology," 55 *Federal Register* 2219, January 23, 1990.

[25] Clinton Administration: Executive Order 12882, "President's Committee of Advisors on Science and Technology," 58 *Federal Register* 62492-62493, November 26, 2003; George W. Bush Administration: Executive Order 13226, "President's Council of Advisors on Science and Technology," 66 *Federal Register* 50523-50524, October 3, 2001; Obama Administration: Executive Order 13539, "President's Council of Advisors on Science and Technology," 75 *Federal Register* 21973-21975, April 27, 2010.

[26] Email from OSTP to CRS, July 27, 2017.

Executive Order 13708 through September 30, 2017.[27] PCAST is an advisory board composed of individuals and representatives from sectors outside the federal government with diverse perspectives and expertise. PCAST advises the President, both directly and through the APST, on science, technology, and innovation policy. In addition, PCAST responds to requests for advice from the National Science and Technology Council. PCAST's members include approximately 20-25 distinguished individuals from industry, education and research institutions, and other organizations outside the federal government. The APST co-chairs PCAST along with one or two other council members. PCAST continues to exist by Executive Order, but members have not yet been appointed during the Trump Administration.[28]

The current executive order gives PCAST a broad remit, stating that its advice "shall include, but not be limited to, policy that affects science, technology, and innovation, as well as scientific and technical information that is needed to inform public policy relating to the economy, energy, environment, public health, national and homeland security, and other topics."[29] PCAST also serves as two statutorily created advisory committees: the President's Innovation and Technology Advisory Committee created by the High Performance Computing Act of 1991 (P.L. 102-194 as amended) and the National Nanotechnology Advisory Panel created by the 21st Century Nanotechnology Research and Development Act (P.L. 108-153).

In 2011, President Obama directed the Department of Energy to provide PCAST with funding and administrative and technical support.[30] Though these functions were transferred to DOE, OSTP has asserted that it continues to exercise policy and programmatic oversight of PCAST

---

[27] Executive Order 13539, "Continuance or Reestablishment of Certain Federal Advisory Committees," 80 *Federal Register* 60271-60273, October 5, 2015.

[28] Email from OSTP to CRS, August 15, 2017.

[29] Executive Order 13539, "President's Council of Advisors on Science and Technology," 75 *Federal Register* 21973- 21975, April 21, 2010.

[30] Executive Order 13596, "Amendments to Executive Orders 12131 and 13539," 76 *Federal Register* 80725-80726, December 27, 2011.

through the APST and PCAST's staff, whose physical office location remains at OSTP.[31]

Budget and Staffing

The PCAST receives no direct appropriations. The OSTP provided funding and support for PCAST through FY2011. In FY2012, the DOE Office of Science assumed this responsibility. According to DOE, it provides support for PCAST staff salary and benefits, travel by committee members, meeting planning support, and other related expenses. Annual funding requested by DOE for PCAST has been under $1 million and has supported up to two FTEs. In FY2017, DOE funding for PCAST is $230,000, a decrease of $311,000 (57.5%) from FY2016. In its FY2018 budget, the Department of Energy states

> The PCAST advisory committee has dissolved and [the DOE Office of Science] is not aware of any plans to reform this committee in FY 2018.[32]

For historical information on DOE appropriations for PCAST, see Table 2.

Table 2. Funding for PCAST ($ in millions)

Fiscal Year	Appropriated
2012	0.6
2013	0.2
2014	0.7
2015	0.8
2016	0.5
2017	0.2

Source: Communication between CRS and Department of Energy Office of Congressional and Intergovernmental Affairs.

---

[31] Email communication from OSTP General Counsel Rachael Leonard to CRS, January 24, 2012.

[32] Department of Energy, *FY2018 Department of Energy's Budget Request to Congress, Volume 4, Science*, p. 339.

Issues and Options for Congress

Certain recurring issues have raised interest among congressional policymakers regarding science and technology policy within the White House. These issues include the titles, roles, and responsibilities of the President's science advisor; the number and policy foci of OSTP Associate Directors; OSTP funding and staffing levels; the participation of OSTP and NSTC in federal agency coordination, priority-setting, and budget allocation; and the stature and influence of PCAST. The following sections address each of these issues.

Title, Rank, Roles, and Responsibilities

Under President Obama, John Holdren served as both OSTP Director and Assistant to the President for Science and Technology (APST). In contrast, under President George W. Bush, John Marburger was given only the title of OSTP Director.[33] Some experts in the S&T community have proposed that the OSTP Director *always* be given the title of APST or be given Cabinet rank. A related issue is whether the roles and responsibilities of the OSTP Director should be undertaken by several appointees rather than one. To a large extent, the appointment of an advisor o a particular position or title arises from presidential discretion. This presidential discretion may limit the ability of Congress to require greater or lesser degrees of access to the President and other key Administration decisionmakers.

Title and Rank

As shown in Appendix A, presidential science advisors have held a variety of titles since the Franklin D. Roosevelt Administration. Of the 13 Administrations reviewed, the most common title has been some variation of Science Advisor to the President (five Administrations), followed by

---

[33] At no time have the positions of OSTP Director and APST been filled by different people.

Special Assistant to the President (four Administrations). The OSTP Director held the title of APST in the Obama, George H. W. Bush, and Clinton Administrations but not in the George W. Bush Administration.[34] The difference between an individual being the OSTP Director and the APST is more than semantic. This section outlines some of the policy issues related to whether the OSTP Director is also designated APST or has Cabinet rank.

Congressional Testimony

Some Members of Congress may wish to have the option to require the individual serving as the President's science advisor to give testimony on OSTP or science and technology policy issues. Others may not place great emphasis on overseeing the role of OSTP Director or APST and may have other sources from which they can obtain S&T analysis and information.

Congress expects that an executive branch official who administers a department or agency established by law will testify before it. This contrasts with an individual whose sole responsibility is to advise the President. Some presidential advisors, such as the OSTP Director, are in units of the EOP established by law and are also subject to confirmation by the Senate. Accordingly, Congress often asks OSTP Directors to testify before it, and may, if necessary, compel them to do so. However, an APST may assert the right not to testify before Congress in accordance with the principles of separation of powers or executive privilege.[35] Some members of the S&T community contend that Congress should permit an individual serving as APST to discriminate between privileged advice to the President that should not be disclosed to Congress and information appropriate to

---

[34] Executive Order 13539, signed by President Obama, specifically designates that the Assistant to the President for Science and Technology shall serve as a co-chair of PCAST, along with one or two of the non-federal members of PCAST. Executive Order 13226, signed by President George W. Bush, stated that the President would designate a "Federal Government official" to serve as a member and co-chair of PCAST. President Bush's designated co-chair was John Marburger, his OSTP Director.

[35] Louis Fisher, "White House Aides Testifying Before Congress," *Presidential Studies Quarterly*, vol. 27, Winter 1997, pp. 140-141. For further discussion, see CRS Report RL31351, *Presidential Advisers' Testimony Before Congressional Committees: An Overview*, by Todd Garvey and Henry B. Hogue.

disclose to Congress.[36] If Congress desires to ensure the availability of the APST for testimony, it might opt to establish the position of APST by statute and require Senate confirmation. Some experts have expressed concern regarding confusion that might arise if Congress could require some Administration staff with "Assistant to the President" titles to testify, but not others.[37] Others have suggested that this might not be an effective approach since, even if such a position were established by statute, a President might opt not to ominate someone for that position or possibly appoint someone to a similarly titled position that does not exist in statute.

Cabinet Rank

Some members of the S&T community have expressed their desire for the OSTP Director to have a greater role and influence in the development of Administration policy. They assert that statutorily designating the OSTP Director as a Cabinet-level position would provide such an enhanced role and influence. In their view, the President would identify an individual nominated for the Cabinet-level OSTP Director position at the same time as other Cabinet members, shortly after the election of a new Administration. If also appointed to serve as APST, the individual could begin work immediately, though exercise of the duties of OSTP Director, with its enhanced stature, would have to await formal nomination and Senate confirmation.[38] If appointed early in a new Administration, some experts in the S&T community contend, the individual filling the APST position could help identify and recruit the best scientists, engineers, health

---

[36] See, for example, Henry Kelly, Ivan Oelrich, Steven Aftergood, and Benn H. Tannenbaum, *Flying Blind: The Rise, Fall and Possible Resurrection of Science Policy Advice in the United States* (Washington, DC: Federation of American Scientists, 2004), http://www.fas.org/pubs/_docs/flying_blind.pdf.

[37] In an email from OSTP General Counsel Rachael Leonard to CRS on January 24, 2012, OSTP stated that "As OSTP Director, Dr. Holdren signed a statement to the Senate Commerce committee prior to his confirmation hearing that he would be available to testify. No APST or OSTP Director/APST has declined to testify."

[38] National Academies, Committee on Science, Engineering, and Public Policy, *Science and Technology for America's Progress: Ensuring the Best Presidential Appointments in a New Administration* (Washington, DC: National Academy Press, 2008), http://www.nap.edu/catalog.php?record_id=12481.

professionals, and other public policy professionals for the S&T policy-related presidential appointments.

Additionally, some contend that an APST/OSTP Director with Cabinet rank would have greater access to the President and other senior Administration staff.[39] They assert that Cabinet rank would enhance the OSTP Director's authority and influence in incorporating scientific and technical viewpoints into Administration decision making. Others contend that the issue of Cabinet rank for the APST/OSTP Director status would be unlikely to substantially improve the APST/OSTP Director's role and influence in EOP activities, including Cabinet meetings.[40]

From a historical perspective, some experts believe that Presidents and their science advisors have unique and idiosyncratic relationships. To these experts, a more important question is how an Administration manages and uses the extensive infrastructure of expert S&T advice that supports all aspects of federal decision making.[41] Scientists, engineers, and S&T policy professionals—both within and outside of the federal government—play a substantial role in providing S&T input to federal policy decision making in areas such as R&D, regulation, procurement, and standards development.

Other experts assert that the organization of the White House determines the S&T advisor's status and access. According to this perspective, if the President relies primarily on a group of White House staff members for advice, the advisor should be the APST. Conversely, if the Cabinet is the primary source of advice, then the science advisor should be made a member of the Cabinet. From this perspective, the title itself is less important than the access to the President that it provides.[42] Other critics contend that rather than focusing on the title, the S&T community

---

[39] National Academies, Committee on Science, Engineering, and Public Policy, *Science and Technology for America's Progress: Ensuring the Best Presidential Appointments in a New Administration* (Washington, DC: National Academy Press, 2008), http://www.nap.edu /catalog.php?record_id=12481.

[40] Based on CRS discussions with Stanley Sokul, George W. Bush Administration Chief of Staff, OSTP, August 14, 2008.

[41] Roger Pielke Jr., "Who Has the Ear of the President?," *Nature* 450:347-348, November 15, 2007, http://www.nature.com/nature/journal/v450/n7168/full/450347a.html.

[42] National Academies, *Science and Technology Advice in the White House: Recommendations for President-Elect George Bush* (Washington, DC: National Academy Press, 1988).

should instead focus on the degree to which an Administration is transparent about its operations.[43]

Roles and Responsibilities

As discussed above, historically OSTP Directors have advised Presidents on S&T policy formulation, R&D budget issues, the policy significance of scientific and technical developments, and STEM education, among other issues. When holding the APST title, the OSTP Director manages the NSTC and co-chairs PCAST.[44] In addition, OSTP Directors can serve as a communication conduit between the EOP and the federal and non-federal S&T community.

One alternative for Congress is to change the current statutory structure and duties of OSTP, separating the various OSTP roles and responsibilities and establishing separate positions and/or organizations for each. For example, the S&T community has debated the utility of having two different individuals serve as APST and OSTP Director. While some believe having two people in these roles might enhance the ability and potential of an APST to be part of the President's inner circle, others believe the potential for conflict between the two is high.[45]

Similarly, some members of the S&T community have suggested that the President appoint co-equal officials, one responsible for science policy and the other for technology policy. Shortly after assuming office, President Obama created the new title of Chief Technology Officer within the EOP and provided it funding through OSTP.[46] The first Chief Technology Officer was also the Associate Director of OSTP for

---

43 For a discussion of this issue, see David Goldston, "US Election: Not the Best Advice," *Nature*, 455:453, September 24, 2008, http://www.nature.com/news/2008/080924/full/455453a.html.

44 President George W. Bush's OSTP Director managed the NSTC and co-chaired PCAST even in the absence of a joint appointment as APST.

45 National Academies, Committee on Science, Engineering, and Public Policy, *Science and Technology in the National Interest: Ensuring the Best Presidential and Federal Advisory Committee Science and Technology Appointments* (Washington, DC: National Academy Press, 2005), http://www.nap.edu/catalog.php?record_id=11152.

46 For more information on the chief technology officer position, see CRS Report R40150, *A Federal Chief Technology Officer in the Obama Administration: Options and Issues for Consideration*, by John F. Sargent Jr.

Technology.[47] Subsequent Chief Technology Officers have not had an Associate Director position. In March 2014, in oral testimony OSTP Director Holdren stated that the Chief Technology Officer did not report to the OSTP Director.[48] Some S&T policy experts have expressed concern that bifurcation of authorities and responsibilities might create conflicts and a lack of integration.[49]

Splitting the functions of OSTP and assigning them to separate individuals or organizations might be challenging due to the size of OSTP's budget and staff.[50] For example, current resources might not effectively support two senior officials and their associated staffs. Congress might opt to increase funding and authorized staffing levels to support such a reorganization.

Number and Policy Foci of OSTP Associate Directors

Current statutory authority provides flexibility to the President with respect to the number of OSTP Associate Directors (up to four) and the scope of their areas of responsibility (entirely at the discretion of the President).[51] President Obama established four Associate Directors with responsibility for discrete policy areas: science; technology and innovation; national security and international affairs; and environment and energy. Under President George W. Bush there were two: an Associate Director for Science and an Associate Director for Technology.

Congress could opt to specify a fixed number of Associate Directors, and could assign some or all of them specific policy foci. Some Members of Congress have undertaken efforts in this regard. For example, in its

---

[47] Aneesh Chopra was the first Chief Technology Officer. Todd Park succeeded him in 2012. Megan Smith succeeded Todd Park in 2014.

[48] Testimony of John Holdren, Director, Office of Science and Technology Policy, Executive Office of the President, The White House, before the Committee on Science, Space, and Technology, March 26, 2014.

[49] David Hatch, "Tech Czar Might Rule Policy Under Obama," *Congressional Daily*, September 10, 2008, http://www.nationaljournal.com/daily/tech-czar-might-rule-policy-under-obama-20080910.

[50] For more information, see "OSTP Budget and Staffing" below.

[51] 42 U.S.C. §6612.

report (S.Rept. 110-124) on the Departments of Commerce and Justice, Science, and Related Agencies Appropriations Act, 2008 (S. 1745, 110th Congress), the Senate Committee on Appropriations recommended that OSTP create the position of Associate Director for Earth Science and Applications to coordinate all federal efforts to better understand and predict changes in the Earth's climate and oceans. Another bill (H.R. 5116, 111th Congress) would have required the OSTP Director to appoint an Associate Director to serve as the Coordinator for Societal Dimensions of Nanotechnology.

In addition, some members of the S&T community have proposed that one or more of the OSTP Associate Director positions should be a joint appointment to the National Economic Council (NEC), National Security Council (NSC), Domestic Policy Council (DPC), or Office of Management and Budget. In this vein, President Obama appointed the OSTP Director and the Chief Technology Officer to the DPC;[52] made OSTP Director Holdren a member of the NEC by providing him with the APST title;[53] added the Chief Technology Officer as a member of the NEC; and issued Presidential Policy Directive 1 (PPD-1) stating that "When science and technology related issues are on the agenda, the NSC's regular attendees will include the Director of the Office of Science and Technology Policy."[54]

OSTP Budget and Staffing

The ability of OSTP to perform its statutory duties depends, in part, on the size of its budget and staff. Figure 2 and Figure 3, above, illustrate

---

[52] White House, *Further Amendments to Executive Order 12859, Establishment of the Domestic Policy Council*, February 5, 2009. For more information, see http://www.whitehouse.gov/the_press_office/Executive-OrderFurtherAmendments-To-Executive-Order-12859Establishment-Of-The-Domestic-Policy-Council/.

[53] White House, *Further Amendments to Executive Order 12835, Establishment of the National Economic Council*, February 5, 2009. For more information, see http://www.whitehouse.gov/the_press_office/Executive-Order-FurtherAmendments-to-Executive-Order-12835-Establishment-of-the-National-Economic-Council/.

[54] Ibid.

OSTP's historical budget and staffing. Between FY1996 and FY2013, the budgets of Presidents Clinton, George W. Bush, and Obama included requests for the authorization of 32-40 full-time equivalent (FTE) positions while the actual number of OSTP-funded staff ranged from 23 to 33. The OSTP has used detailees and fellows to supplement its core staffing. During the George W. Bush Administration, detailees and fellows provided approximately half of OSTP's total staff; during the Clinton Administration, detailees and fellows accounted for approximately two-thirds of total OSTP staff; toward the end of the Obama Administration, detailees, fellows, and IPAs account for approximately two-thirds of total OSTP staff.

Some in the S&T community have expressed concerns that OSTP needs to have more career civil service professional staff and a larger budget.[55] In their view, additional career staff, who would continue to serve from one presidential Administration to the next, would help maintain institutional knowledge and provide a solid understanding of government operations. More career staff might also enable a new Administration to move more quickly on S&T policy issues and provide enhanced support to political appointees during presidential transitions. Reports expressing these views assert that this change would make OSTP staff similar to other EOP expert staff, such as those employed at OMB.[56]

Additional funding, these reports assert, would also provide OSTP with sufficient staff to conduct special analyses on emerging issues. Currently, such analyses are generally provided by OSTP's federally funded research and development center (FFRDC), the Science and Technology Policy Institute (STPI). (See "Science and Technology Policy Institute" box, above.)

---

[55] Henry Kelly, Ivan Oelrich, Steven Aftergood, and Benn H. Tannenbaum, *Flying Blind: The Rise, Fall and Possible Resurrection of Science Policy Advice in the United States* (Washington, DC: Federation of American Scientists, 2004), http://www.fas.org/pubs/_docs/flying_blind.pdf; and Jennifer Sue Bond, Mark Schaefer, David Rejeski, Rodney W. Nichols, *OSTP 2.0: Critical Upgrade: Enhancing Capacity for White House Science and Technology Policymaking: Recommendations for the Next President* (Washington, DC: Woodrow Wilson International Center for Scholars, June 2008).

[56] According to the FY2015 budget request, the OMB FY2014 budget was $89.3 million, which supported 470 full time equivalent staff. For more information, see http://www.whitehouse.gov/sites/default/files/docs/2015-eopbudget_03132014.pdf.

Congress may wish to maintain the current staffing approach. Should Congress wish to enhance the funding and staffing of OSTP, it can do so through the appropriations process. The OSTP received $5.5 million for FY2017. For funding levels in previous years, see Figure 2 and Appendix B. During the Obama Administration, funding ranged from $4.5 million (in FY2012) to $7.0 million (in FY2010).

OSTP and NSTC Participation in Federal Agency Coordination, Priority-Setting, and Budget Allocation

OSTP and the NSTC participate in coordinating, setting priorities for, and allocating the budget for federal S&T activities. S&T policy organizations have suggested enhancing this participation. The following sections address OSTP interactions with other EOP offices and the science community, the role of the Director of OSTP, and the role of the NSTC.

OSTP Interactions with Other EOP Offices and the Science Community

Policy tensions and power struggles between OSTP and other EOP offices and between presidential Administrations and the science community are not new. During the George H. W. Bush Administration, tension existed between OSTP Director D. Allan Bromley and other high-ranking White House officials over the extent of Administration support for federal funding of commercial technology development.[57] In July 1981, George Keyworth, Reagan Administration OSTP Director, stirred controversy in the science community on his first speech to the American Association for the Advancement of Science (AAAS) by asserting that "Nowhere is it indicated that the OSTP or its director is to represent the interests of the scientific community as a constituency."[58] Carter

---

[57] Bob Davis, "White House, Reversing Policy Under Pressure, Begins to Pick High-Tech Winners and Losers," *Wall Street Journal*, May 13, 1991, p. A16; Bob Davis, "White House Tries to Distance Itself from Panel Report," *Wall Street Journal*, April 26, 1991, p. A16.

[58] Barbara J. Culliton, "Keyworth Gives First Speech," *Science*, July 7, 1981, pp. 183-184.

Administration OSTP Director Frank Press battled the Council on Environmental Quality (CEQ), opposing the CEQ-advocated use of federal subsidies to the then-infant solar power industry and instead supporting a balance between market demand and scientific discovery.[59]

Role of OSTP Director

Some reports from the S&T community suggest that the OSTP Director should take a greater role in coordination, priority-setting, and budget allocation regarding the federal R&D budget;[60] energy;[61] STEM education;[62] international S&T policy;[63] and federal-state S&T policy.[64] In addition, some members of the S&T policy community have suggested that the OSTP Director play a greater role in EOP policy bodies involved in priority-setting and budget allocation, such as the OMB, NEC, CEQ, DPC, and NSC.[65] For example, Congress could require the OSTP Director to play a greater role (e.g., certification of priorities or budgets) in setting priorities at the federal agencies, particularly for multi-agency and inter-agency activities.

---

[59] David Dickson, *The New Politics of Science* (NY: Pantheon Books/Random House, Inc., 1984), pp. 37-38.

[60] Henry Kelly, Ivan Oelrich, Steven Aftergood, and Benn H. Tannenbaum, *Flying Blind: The Rise, Fall and Possible Resurrection of Science Policy Advice in the United States* (Washington, DC: Federation of American Scientists, 2004), http://www.fas.org/pubs/_docs/flying_blind.pdf.

[61] Senator Jeff Bingaman, "The Energy Challenge We Face and the Strategies We Need," The Karl Taylor Compton Lecture, Massachusetts Institute of Technology, April 25, 2008.

[62] National Science Board, *National Action Plan for Addressing the Critical Needs of the U.S. Science, Technology, and Mathematics Education System* (Ballston, VA: National Science Foundation, 2007), http://www.nsf.gov/nsb/ documents/2007/stem_action.pdf.

[63] National Science Board, *International Science and Engineering Partnerships: A Priority for U.S. Foreign Policy and Our Nation's Innovation Enterprise*, NSB 08-4 (Arlington, VA: National Science Foundation, 2008), http://www.nsf.gov/nsb/publications/2008/nsb084.pdf. Jennifer Sue Bond, Mark Schaefer, David Rejeski, Rodney W. Nichols, *OSTP 2.0: Critical Upgrade: Enhancing Capacity for White House Science and Technology Policymaking: Recommendations for the Next President* (Washington, DC: Woodrow Wilson International Center for Scholars, June 2008).

[64] Jennifer Sue Bond, Mark Schaefer, David Rejeski, Rodney W. Nichols, *OSTP 2.0: Critical Upgrade: Enhancing Capacity for White House Science and Technology Policymaking: Recommendations for the Next President* (Washington, DC: Woodrow Wilson International Center for Scholars, June 2008).

[65] Ibid.

Role of NSTC

Another recommendation found in these S&T community reports is to make the NSTC's authority equivalent to that of the NSC.[66] The NSTC, they assert, lacks the influence of NSC. The differences in statutory authority, staff, and budget are among the reasons cited for this disparity.

The NSTC has participated in presidential decision-making processes in different ways in different Administrations. For example, during the Clinton Administration, the NSTC issued six Presidential Review Directives (PRDs). The PRDs served as the basis for gathering information and policy options for the President. President Clinton then had this information available as he developed eight Presidential Decision Directives (PDDs) establishing new policy.[67] The NSTC has not developed PRDs or their equivalents since the end of the Clinton Administration.

Some experts in the S&T community suggest that the NSTC should issue formal directives rather than contributing input and deliberations into the policy documents of other entities. These experts argue that contributing input to and deliberating on other entity policy documents puts S&T and the NSTC in a supportive role. These experts assert that, in some situations, S&T input and ramifications should have a more prominent influence on public policy.[68]

In 2012, the Obama Administration asserted that it had undertaken efforts to revitalize and streamline the efforts of the NSTC. The Administration cited its establishment of a fifth NSTC committee—the Committee on Science, Technology, Engineering, and Math (STEM) Education—to coordinate federal programs and activities in support of STEM education. The Administration stated that under President Obama NSTC committees met two or three times annually and each subcommittee

---

[66] Henry Kelly, Ivan Oelrich, Steven Aftergood, and Benn H. Tannenbaum, *Flying Blind: The Rise, Fall and Possible Resurrection of Science Policy Advice in the United States* (Washington, DC: Federation of American Scientists, 2004) at http://www.fas.org/pubs/_docs/flying_blind.pdf.

[67] A list is available at http://www.fas.org/irp/offdocs/direct.htm.

[68] Henry Kelly, Ivan Oelrich, Steven Aftergood, and Benn H. Tannenbaum, *Flying Blind: The Rise, Fall and Possible Resurrection of Science Policy Advice in the United States* (Washington, DC: Federation of American Scientists, 2004) at http://www.fas.org/pubs/_docs/flying_blind.pdf.

met at least quarterly. The Obama Administration also asserted that it "oversaw the restructuring of the original NSTC committees, with elimination of interagency efforts, where appropriate, and initiation of new efforts, as indicated by Administration priorities and/or Congressional mandates."[69]

Options for Congress

Congress might choose to leave the roles of the OSTP Director and the NSTC in the budget process unchanged, might choose to increase their authorities, might choose to increase its oversight of their roles, or might do a combination of these.

Congress might mandate that OSTP review the S&T components of agency budgets prior to submission to OMB and empower OSTP to alter the distribution of funding between S&T priorities based on their relative importance. Such authority might increase the ability of OSTP to harmonize and coordinate S&T expenditures among federal agencies. Federal agencies might resist such a change in authority, as it might further complicate the budget development and submission process and create competition between OSTP and OMB directives. In addition, such a mandate might have unintended consequences. For example, agencies might not choose to identify S&T-related programs to evade the mandate.

Congress might require that NSTC or OSTP review the S&T components of agency budgets to assess the correspondence between NSTC multi-agency R&D strategies and proposed federal investments. A hallmark of multi-agency R&D investment is the need to coordinate the magnitude and mission goals of agency investments in order to achieve broader federal R&D goals. Such a review might increase transparency regarding progress towards these broader federal R&D goals, but it might also require increases in expenditures. Identifying cross-cutting funding and efforts might require dedicated program offices and staff to track and report on multi-agency activities.

---

[69] Email from OSTP General Counsel Rachael Leonard to CRS, January 24, 2012.

Congress might choose to formalize the NSTC structure and organization and provide additional funding and personnel to increase the robustness of its process. Providing statutory underpinnings for the NSTC might enable Congress to obtain greater insight into the activities of the NSTC through reporting requirements and oversight of its activities. Alternatively, Congress could mandate that the OSTP Director provide regular reports on the activities of the NSTC.

Stature and Influence of PCAST

As discussed above, PCAST advises the President on science, technology, and innovation-related issues. PCAST's members include individuals from industry, education and research institutions, and other organizations outside the federal government.

Legislative activity has focused less on PCAST than on the NSTC. In a 2008 report, some experts in the S&T policy community asserted that the stature and influence of PCAST had declined as PCAST focused on a narrower set of issues less likely to garner presidential interest.[70] These experts noted that although President George H. W. Bush held the first PCAST meeting at Camp David and participated in PCAST meetings, Presidents Clinton and George W. Bush only met occasionally for short periods of time with PCAST chair or committee members.

According to OSTP, PCAST co-chairs met with President Obama and senior EOP officials several times for focused discussions on specific topics that PCAST should undertake for its studies, updates on studies in progress, briefings on completed studies prior to public release, and actions the President could consider in response to PCAST's recommendations.[71]

As a federal advisory committee, PCAST is unusual in that Executive Order 13539 directs that it is to be co-chaired by the APST and one of its

---

[70] Center for the Study of the Presidency, Study Group on Presidential Science and Technology Personnel Advisory Assets, "*Presidential Leadership to Ensure Science and Technology in Service of National Needs: A Report to the 2008 Candidates,*" Summer 2008.

[71] Email communication from OSTP General Counsel Rachael Leonard to CRS, January 24, 2012.

members, as opposed to having an independent chair not directly associated with the Administration. Federal advisory committees generally do not have Administration staff as chairs. Administration staff are more commonly included as ex-officio members.[72] The designation of the APST as co-chair may reduce PCAST's ability to provide independent thinking to the White House and may place the APST in an awkward position if PCAST members disagree with White House policy. Alternatively, PCAST recommendations may be more likely to be acted upon if the co-chair role of the APST helps to inform PCAST deliberations of Administration perspectives.

Some S&T policy organizations have suggested strengthening PCAST by broadening its mandate, explicitly including national and homeland security issues within its remit, enhancing its independence, and increasing its staff significantly.[73] Other suggestions include selecting the chair of PCAST solely from its non-Administration members; appointing members to staggered, overlapping terms unrelated to presidential and congressional election cycles; and providing all members with security clearances. President Obama authorized the APST to

> request that members of the PCAST, its standing subcommittees, or ad hoc groups who do not hold a current clearance for access to classified information, receive security clearance and access determinations

---

[72] For example, the Director of the National Science Foundation is an ex-officio member of the National Science Board and the charter of the National Science Advisory Board for Biosecurity allows for non-voting ex-officio representatives of the Executive Office of the President and a number of federal agencies and entities. For more information, see CRS Report R40520, *Federal Advisory Committees: An Overview*, by Wendy Ginsberg (out of print; availabe from the author).

[73] See for example, Carnegie Commission on Science, Technology, and Government, *Science & Technology and the President* (New York: Carnegie Corporation of New York, October 1988); Henry Kelly, Ivan Oelrich, Steven Aftergood, and Benn H. Tannenbaum, *Flying Blind: The Rise, Fall and Possible Resurrection of Science Policy Advice in the United States* (Washington, DC: Federation of American Scientists, 2004); and Center for the Study of the Presidency, Study Group on Presidential Science and Technology Personnel Advisory Assets, *Presidential Leadership to Ensure Science and Technology in Service of National Needs: A Report to the 2008 Candidates,* Summer 2008.

pursuant to Executive Order 12968 of August 2, 1995, as amended, or any successor order.[74]

OSTP asserted that most of the PCAST members had obtained security clearances so that PCAST could undertake studies related to national security.[75]

Some experts in the S&T community have also suggested increasing the number of presidential advisory committees. For example, they propose advisory committees focused on specific S&T policy issues, such as a Federal-State Science and Technology Council to enhance dialogue with the states, particularly on STEM education.[76] The costs of establishing such new advisory committees may pose a challenge to their creation. In addition, requirements of the Federal Advisory Committee Act (P.L. 92-463) regarding justification of any new advisory committee, its membership, and associated ethics rules (including financial disclosure) may complicate the establishment of new committees and the recruitment of committee members. As noted above, PCAST has taken on the responsibilities of several topic-specific advisory committees established in statute.

If Congress wanted the President to establish additional presidential advisory committees—either to address areas not currently covered by PCAST or to address issues currently covered by PCAST but with separate committees focused on a particular area (e.g., nanotechnology, networking and information technology)—it might opt to provide additional funding to OSTP expressly for this purpose.

---

[74] Executive Order 13539, "President's Council of Advisors on Science and Technology," April 21, 2010, http://www.gpo.gov/fdsys/pkg/FR-2010-04-27/pdf/2010-9796.pdf.

[75] Email communication from OSTP General Counsel Rachael Leonard to CRS, January 24, 2012.

[76] Jennifer Sue Bond, Mark Schaefer, David Rejeski, Rodney W. Nichols, *OSTP 2.0: Critical Upgrade: Enhancing Capacity for White House Science and Technology Policymaking: Recommendations for the Next President* (Washington, DC: Woodrow Wilson International Center for Scholars, June 2008); and Center for the Study of the Presidency, Study Group on Presidential Science and Technology Personnel Advisory Assets, *Presidential Leadership to Ensure Science and Technology in Service of National Needs: A Report to the 2008 Candidates*, Summer 2008.

In 2012, OSTP asserted that President Obama had increased the role and influence of PCAST by considering and taking action on PCAST recommendations, including

- funding a new influenza vaccine manufacturing improvement initiative to shorten the time frame for production of pandemic influenza vaccines, including dedication of the first U.S. cell-based influenza vaccine plant;
- proposing preparation of an additional 100,000 K-12 STEM teachers by the end of the decade and establishment of an Advanced Research Projects Agency-Education (ARPA-ED);
- accelerating adoption of Electronic Health Records and developing standards for health information exchange over the Internet, and metadata for Stages 2 and 3 of the electronic health records meaningful use criteria;
- establishing the Advanced Manufacturing Partnership, including initial funding for new initiatives; and
- undertaking a Quadrennial Technology Review at the Department of Energy.[77]

In 2012, OSTP asserted that during the Obama Administration PCAST had met six times per year compared to three or four times per year during the George W. Bush Administration. In addition, OSTP asserted in 2012 that PCAST had "met with every major Administration leader in science and technology, including Cabinet-level Secretaries, to gather their views on the topics most useful for PCAST to address, and to discuss implementation of PCAST's recommendations."[78]

In addition, OSTP has stated that the Obama Administration provided PCAST with the staff and financial resources necessary to develop reports in a timely fashion for Congress and the Administration. These resources, according to OSTP at the time, increased the ability of PCAST to provide

---

[77] Email communication from OSTP General Counsel Rachael Leonard to CRS, January 24, 2012.

[78] Ibid.

reports and recommendations. PCAST released 18 reports during the eight years of the Bush Administration; through the first six years of the Obama Administration, PCAST had released 26 reports through January 2015.[79]

APPENDIX A. PRESIDENT'S SCIENCE AND TECHNOLOGY POLICY ADVISORS

Table A-1. President's Science and Technology Policy Advisors and Predecessor Organizations to OSTP, NSTC, and PCAST, 1941-Present

President	Advisors with Title(s) (Years in Office)	Executive Office of the President Agency (Year Established)	Interagency Coordination Organization[a] (Year Established)	Advisory Committee (Year Established)
F.D. Roosevelt	Vannevar Bush[b] (1941-1945), Director, Office of Scientific Research and Development	Office of Scientific Research and Development (OSRD; 1941)		Science Advisory Board (1933)
Truman	John Steelman[b] (1946-1947), Special Assistant to the President (1945-1946); Assistant to the President (1946-1953); Chairman, The President's Scientific Research Board (1946-1947) Oliver Buckley[b] (1951-1952), Chair, Science Advisory Committee (SAC) Lee DuBridge[b] (1952-1953), Chair, SAC		The President's Scientific Research Board (1946-1947);[c] Interdepartmental Committee for Scientific Research (1947)[c]	Science Advisory Committee (SAC) of the Office of Defense Mobilization (1946)[c]

[79] http://www.whitehouse.gov/administration/eop/ostp/pcast/docsreports.

President	Advisors with Title(s) (Years in Office)	Executive Office of the President Agency (Year Established)	Interagency Coordination Organization[a] (Year Established)	Advisory Committee (Year Established)
Eisenhower	Lee DuBridge[b] (1953-1956), Chair, SAC; Science Advisor to the President Isidor I. Rabi[b] (1956-1957), Chair, SAC; Science Advisor to the President James Killian Jr. (1957-1959), Special Assistant to the President for Science and Technology; Chair, President's Science Advisory Committee (PSAC) George Kistiakowsky (1959-1961), Special Assistant to the President for Science and Technology; Chair, PSAC	Office of the Special Assistant to the President for Science and Technology (1957)	Federal Council for Science and Technology (FCST) (1959)	SAC (1953-56); President's Science Advisory Committee (PSAC; 1957, replaced SAC).
Kennedy	Jerome Wiesner (1961-1963), Special Assistant to the President for Science and Technology; Director, OST; Chair, FCST; Chair, PSAC	Office of Science and Technology (OST; 1962)	FCST	PSAC
Johnson	Jerome Wiesner (1963-1964), Special Assistant to the President for Science and Technology; Director, OST; Chair, FCST; Chair, PSAC Donald Hornig (1964-1969), Special Assistant to the President for Science and Technology; Director, OST; Chair, FCST: Chair, PSAC	OST	FCST	PSAC

Table A-1. (Continued)

President	Advisors with Title(s) (Years in Office)	Executive Office of the President Agency (Year Established)	Interagency Coordination Organization[a] (Year Established)	Advisory Committee (Year Established)
Nixon	Lee DuBridge (1969-1970), Science Advisor to the President; Director, OST Edward David Jr. (1970-1973), Science Advisor to the President; Director, OST H. Guyford Stever (1973-1974), Science Advisor to the President; Chair, FCST	OST (until 1973, when office abolished)[d]	FCST	PSAC (until 1973, when member resignations were accepted and no new appointments were made).
Ford	H. Guyford Stever (1974-1977); Science Advisor to the President; Director, Office of Science and Technology Policy (OSTP)	Office of Science and Technology Policy (1976)	Federal Coordinating Council for Science, Engineering, and Technology (FCCSET; 1976, replaced FCST)	Intergovernmental Science, Engineering, and Technology Panel (ISETAP; 1976);[e] President's Council on Science and Technology (PCST; 1976)
Carter	Frank Press (1977-1981); Science and Technology Advisor to the President; Director, OSTP; Chair, FCCSET	OSTP	FCCSET dissolved as statutory entity and reestablished under an executive order (1978)	PCST (until 1978, abolished with its functions transferred to President by executive order); ISETAP (until 1978, dissolved as statutory entity and reestablished under an executive order)

President	Advisors with Title(s) (Years in Office)	Executive Office of the President Agency (Year Established)	Interagency Coordination Organization[a] (Year Established)	Advisory Committee (Year Established)
Reagan	George Keyworth II (1981-1985), Science Advisor to the President; Director, OSTP William R. Graham (1986-1989), Science Advisor to the President; Director, OSTP	OSTP	FCCSET	White House Science Council (1982; reports to Science Advisor, not President; established by Science Advisor, not executive order)
G.H.W. Bush	D. Allan Bromley (1989-1993), Assistant to the President for Science and Technology; Director, OSTP; Chair, PCAST	OSTP	FCCSET	President's Council of Advisors on Science and Technology (PCAST; 1990)
Clinton	John Gibbons (1993-1998), Assistant to the President for Science and Technology; Director, OSTP; Co-Chair, PCAST	OSTP	National Science and Technology Council (NSTC; 1993)	PCAST (Name changed to President's Committee of Advisors on Science and Technology; 1993)
	Neal Lane (1998-2001), Assistant to the President for Science and Technology; Director, OSTP; Co-Chair, PCAST			
G.W. Bush	John Marburger, III (2001-2009), Science Advisor to the President; Director, OSTP; Co-Chair, PCAST	OSTP	NSTC	PCAST (Name changed back to President's Council of Advisors on Science and Technology; 2001)

President	Advisors with Title(s) (Years in Office)	Executive Office of the President Agency (Year Established)	Interagency Coordination Organization[a] (Year Established)	Advisory Committee (Year Established)
Obama	John P. Holdren (2009-2017), Assistant to the President for Science and Technology; Director, OSTP; Co-Chair, PCAST	OSTP	NSTC	PCAST
Trump	No appointment or nomination had been made as of August 2017.	OSTP	NSTC	PCAST

Sources: Congressional Research Service, based on information from the following sources: Public Papers of the Presidents (Washington, DC: GPO) with the following volumes were used as references: Dwight D. Eisenhower (1957, 1960); Lyndon B. Johnson (1962, 1966, 1967); Richard M. Nixon (1969, 1970, 1973), Gerald Ford (1976- 1977), Jimmy Carter (1977, 1978), Ronald Reagan (1981, 1983, 1986), and George H. W. Bush (1989); Jeffrey K. Stine, "A History of Science Policy in the United States, 1940-1985," Report for the House Committee on Science and Technology Task Force on Science Policy, 99th Congress, 2nd session, Committee Print (Washington, DC: GPO, 1986), available at http://ia341018.us.archive.org/2/items/historyofscience00unit/historyofscience00unit.pdf; William T. Golden (ed.), *Science Advice to the President* (New York: Pergamon Press, 1979); William G. Wells, "Science Advice and the Presidency: 1933-1976," Dissertation, School of Government and Business Administration (Washington, DC: George Washington University, 1977); OSTP, "Previous Science Advisors," website at http://www.whitehouse.gov/administration/eop/ostp/about/leadershipstaff/previous; Truman Library at http://www.trumanlibrary.org/hstpaper/steelman.htm; "Lee Alvin DuBridge (Part II) (1901-1993), Interviewed by Judith R. Goodstein," Oral History, February 20, 1981, California Institute of Technology Archives at http://oralhistories.library.caltech.edu/68/01/OH_DuBridge_2.pdf; Nixon Presidential Library Archives, Officials of Administration at http://nixon.archives.gov/thelife/apolitician/thepresident/officialsofadministration.php; John T. Woolley and Gerhard Peters, The American Presidency Project [online], Santa Barbara, CA: University of California (hosted), Gerhard Peters (database) at http://www.presidency.ucsb.edu/; National Archives, "Records of the Office of Science and Technology," web page at http://www.archives.gov/research/guide-fed-records/ groups/359.html. Other sources include Executive Order 9912, "Establishing the Interdepartmental Committee on Scientific Research and Development," 12 *Federal Register* 8799, December 27, 1947, at http://www.presidency.ucsb.edu/ws/index.php?pid=60725; Executive Order 9913, "Terminating the Office of Scientific Research and Development and Providing for the Completion of its Liquidation," 12 *Federal Register* 8799, December 27, 1947, at http://www.presidency.ucsb.edu/ws/index.php?pid= 78155; Executive Order 10807, "Federal Council for Science and Technology, 24 *Federal Register* 1897, March 17, 1959; Executive Order 12039, "Relating to the

Transfer of Certain Science and Technology Policy Functions," 43 *Federal Register* 8095; February 28, 1978 at http://www.presidency.ucsb.edu/ws/index.php?pid=30416; Executive Order 12881, "Establishment of the National Science and Technology Council," 58 *Federal Register* 226, November 23, 1993, p. 62491, at http://www.archives.gov/federal-register/ executive-orders/pdf/12881.pdf; Executive Order 12882, "Executive Order 12882—President's Committee of Advisors on Science and Technology," 58 *Federal Register* 226, November 26, 1993, p. 62493, at http://www.archives.gov/federal-register/executive-orders/pdf/12882.pdf; Executive Order 13226, "President's Council of Advisors on Science and Technology," 66 *Federal Register* 192, October 3, 2001, pp. 50523-52524, at http://frwebgate.access.gpo.gov/cgi-bin/ getdoc.cgi?dbname=2001_register&docid=fr03oc01-141.pdf; Executive Order 13539, "President's Council of Advisors on Science and Technology," 75 *Federal Register* 21973-21975, April 27, 2010, http://edocket.access.gpo.gov/2010/pdf/2010-9796.pdf; U.S. President (Kennedy), "Special Message to the Congress Transmitting Reorganization Plan 2 of 1962," Public Papers of the Presidents of the United States: John F. Kennedy, 1962, March 29, 1962, at http://www.presidency.ucsb.edu/ws/index.php?pid=24601&st=Reorganization+Plan+No.+2+of+ 1962&st1=; U.S. President (Nixon), "Message to the Congress Transmitting Reorganization Plan 1 of 1973 Restructuring the Executive Office of the President," Public Papers of the Presidents of the United States: Richard M. Nixon, January 26, 1973, at http://www.presidency.ucsb.edu/ ws/index.php?pid= 3819&st=Reorganization+Plan+No.+1+of+1973&st1=; U.S. President (Carter), "Executive Office of the President Message to the Congress Transmitting Reorganization Plan No. I of 1977," Public Papers of the Presidents of the United States: Jimmy Carter, July 15, 1977, at http://www.presidency.ucsb.edu/ws/index.php?pid=7809&st= Reorganization+Plan+No.+1+of+1977&st1=.

Notes: The science advisors may have additional titles not represented in this table. In recent times, the hierarchy of assistants to the President within the White House Office is as follows, going from high to low: Assistant to the President, Deputy Assistant to the President, Special Assistant to the President. (National Archives and Records Administration, The United States Government Manual 2007-2008 (Washington, DC: GPO, 2007) at http://www.gpoaccess.gov/gmanual/ browse-gm-07.html.)

[a] Prior to the designation of any individual to serve as the President's science and technology advisor, President Theodore Roosevelt appointed the Committee on the Organization of Scientific Work to assess the central organization of government scientific bureaus (agencies) with a focus on eliminating duplication.

[b] Opinions differ on who is the first presidential science advisor. The OSTP website states that Oliver Buckley was the first science advisor; it does not include either Vannevar Bush or John Steelman in its list of presidential science advisors (source: OSTP, "Previous Science Advisors," http://www.whitehouse.gov/administration/eop/ ostp/about/leadershipstaff/previous, accessed February 2, 2015). Others believe the latter two individuals were presidential science advisors as well. As OSRD Director, Vannevar Bush, submitted a report, *Science: The Endless Frontier*, to the President Franklin Roosevelt Administration that is the foundation for today's federal S&T policy. President Truman asked that John Steelman, as Director of War Mobilization and Reconversion in the EOP, chair a Presidential Scientific Research Board that was to make recommendations on how to enhance coordination and efficiency of federal R&D. Once this report was released, President Truman asked Steelman, a Presidential Assistant, to act as a liaison between the President and the newly formed Interdepartmental Committee on Scientific Research and Development. Buckley, Lee DuBridge, and Isidor Rabi were all Chairs of the Science Advisory Committee and as such, were given the title of Presidential science advisors. For more

discussion of this issue, see "Oral History Interview with William T. Golden" at http://www.trumanlibrary.org/oralhist/goldenw.htm.

[c] For an understanding of the charges to the different scientific advisory boards and committees, see "Letter to the Chairman, Science Advisory Committee" at http://trumanlibrary.org/publicpapers/viewpapers.php?pid=301; executive order establishing the President's Scientific Research Board, available at http://www.trumanlibrary.org/executiveorders/index.php?pid=467; and the Interdepartmental Committee for Scientific Research, available at http://www.trumanlibrary.org/publicpapers/index.php?pid=1847&st=&st1=.

[d] On January 26, 1973, as part of a reorganization plan, the Office of Science and Technology within the Executive Office of the President was abolished. All of its duties, including that of Science Advisor, were transferred to the National Science Foundation (NSF). As a result, the NSF Director became the Science Advisor. For more details, see http://www.presidency.ucsb.edu/ws/index.php?pid=3819&st=&stl=.

[e] ISETAP members included the OSTP Director, NSF Director, and state, local, and regional officials.

APPENDIX B. HISTORICAL OSTP FUNDING

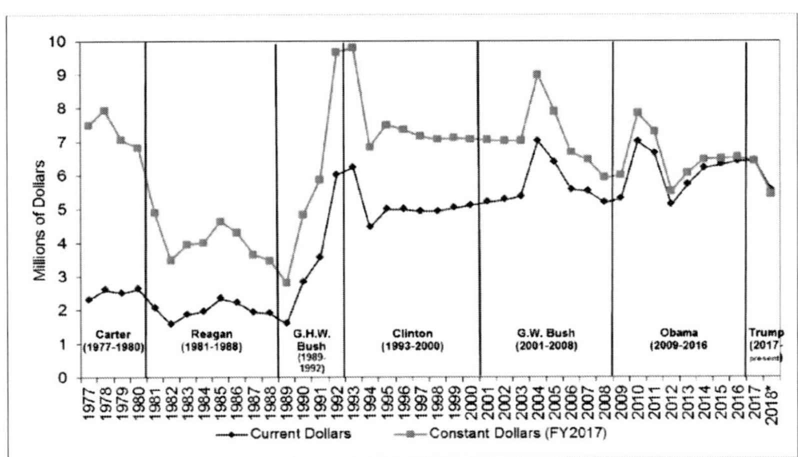

Sources: Congressional Research Service. Data from OMB Public Budget Database; budget requests; and congressional appropriations acts and committee reports, FY1 977-FY20 18; PCAST funding data from the Department of Energy.

Notes: In FY2008, Congress directed NSF to transfer $2.240 million to OSTP for Science and Technology Policy Institute (STPI) (not shown). If the STPI funding were included, FY2008 funding for OSTP would be $7.424 million in current dollars. The data above includes in funding for PCAST provided by the Department of Energy starting in FY2012. Funding in FY20 13 is post-sequestration.

* = request level.

Figure B-1. OSTP Funding, FY1977-FY2018.

In: Science Policies and Programs
Editor: Johnnie Rodgers
ISBN: 978-1-53614-107-8

Chapter 6

SCIENCE AND TECHNOLOGY ISSUES IN THE 115TH CONGRESS*

Frank Gottron

ABSTRACT

Science and technology (S&T) have a pervasive influence over a wide range of issues confronting the nation. Public and private research and development spur scientific and technological advancement. Such advances can drive economic growth, help address national priorities, and improve health and quality of life. The constantly changing nature and ubiquity of science and technology frequently create public policy issues of congressional interest.

The federal government supports scientific and technological advancement directly by funding and performing research and development and indirectly by creating and maintaining policies that encourage private sector efforts. Additionally, the federal government establishes and enforces regulatory frameworks governing many aspects of S&T activities.

---

* This is an edited, reformatted, and augmented version of a Congressional Research Service report, R44786, dated May 23, 2017.

This chapter briefly outlines an array of science and technology policy issues that may come before the 115th Congress. Given the rapid pace of S&T advancement and its importance in many diverse public policy issues, S&T-related issues not discussed in this chapter may come before the 115th Congress. The selected issues are grouped into 9 categories:

- Overarching S&T Policy Issues,
- Agriculture,
- Biomedical Research and Development,
- Defense,
- Energy,
- Environment and Natural Resources,
- Homeland Security,
- Information Technology,
- Physical and Material Sciences, and
- Space.

Each of these categories includes concise analysis of multiple policy issues. The material presented in this chapter should be viewed as illustrative rather than comprehensive. Each section identifies CRS reports, when available, and the appropriate CRS experts to contact for further information.

INTRODUCTION

Science and technology (S&T) play an increasingly important role in our society. Advances in science and technology can help drive economic growth, improve human health, increase agricultural productivity, and meet national priorities.

Federal policies affect scientific and technological advancement on several levels. The federal government directly funds research and development (R&D) activities to achieve national goals or support national priorities, such as funding basic life science research through the National Institutes of Health (NIH) or developing new weapons systems in the Department of Defense (DOD). The federal government establishes and maintains the legal and regulatory framework that affects S&T activities in the private sector. Federal tax, trade, intellectual property, regulatory, and education policies can have large effects on private sector S&T activity.

This chapter serves as a brief introduction to many of the science and technology policy issues that may come before the 115th Congress. Each issue section provides background information and outlines selected policy issues that may be considered. Each issue includes a heading entitled "For Further Information" that provides the author's contact information and the titles of relevant CRS reports containing more detailed policy analysis and information.

OVERARCHING S&T POLICY ISSUES

Several issues of potential congressional interest apply to federal science and technology policy in general. This section begins with a brief introduction to the roles each branch of the federal government plays in S&T policymaking, then discusses overall federal funding of research and development. Additional sections address issues related to the America COMPETES Act; oversight of federally supported academic research; technology transfer; the adequacy of the science and engineering workforce; science, technology, engineering, and mathematics (STEM) education; and innovation-related tax policy.

Federal Science and Technology Policymaking Enterprise

The federal S&T policymaking enterprise is composed of an extensive and diverse array of stakeholders in the executive, legislative, and judicial branches. The enterprise fosters, among other things, the advancement of scientific and technical knowledge; STEM education; the application of S&T to achieve economic, national security, and other societal benefits; and the use of S&T to improve federal decisionmaking.

Federal responsibilities for S&T policymaking are highly decentralized. In addition to appropriating funding for S&T programs,

Congress enacts laws to establish, refine, and eliminate programs, policies, regulations, regulatory agencies, and regulatory processes that rely on S&T data and analysis. However, Congress's authorities related to S&T policymaking are diffuse. Many House and Senate committees have jurisdiction over important elements of S&T policy. In addition, there are dozens of informal congressional caucuses in areas of S&T policy such as research and development, specific S&T disciplines, and STEM education.

The President formulates annual budgets, policies, and programs for consideration by Congress; issues executive orders and directives; and directs the executive branch departments and agencies responsible for implementing S&T policies and programs. The Office of Science and Technology Policy, in the Executive Office of the President, advises the President and other Administration officials on S&T issues.

Executive agency responsibilities for S&T policymaking are also diffuse. Some agencies have broad S&T responsibilities (e.g., the National Science Foundation). Others use S&T to meet a specific federal mission (e.g., defense, energy, health, space). Regulatory agencies have S&T responsibilities in areas such as nuclear energy, food and drug safety, and environmental protection.

Federal court cases and decisions often affect U.S. S&T policy. Decisions can have an impact on the development of S&T (e.g., decisions regarding the U.S. patent system); S&T-intensive industries (e.g., the break-up of AT&T in the 1980s); and the admissibility of S&T-related evidence (e.g., DNA evidence).

Federal Funding for Research and Development

The federal government has long supported the advancement of scientific knowledge and technological development through investments in R&D. Federal R&D funding seeks to address a broad range of national interests, including national defense, health, safety, the environment, and energy security; advance knowledge generally; develop the scientific and engineering workforce; and strengthen U.S. innovation and competitiveness. The federal government has played an important role in supporting R&D efforts which have led to scientific breakthroughs and new technologies, from jet aircraft and the Internet to communications satellites and defenses against disease.

Between FY2009 and FY2016, federal R&D funding fell from $147.3 billion to $146.1 billion, a reduction of $1.2 billion (0.8% in current dollars, 11.1% in constant dollars); funding has rebounded from a period low of $130.3 billion in FY2013. The decline was a reversal of sustained growth in federal R&D funding for more than half a century, and has stirred debate about the potential long-term effects on U.S. technological leadership, innovation, competitiveness, economic growth, and job creation. Concerns about reductions in federal R&D funding have been exacerbated by increases in the R&D investments of other nations (China, in particular); globalization of R&D and manufacturing activities; and trade deficits in advanced technology products, an area in which the United States previously ran trade surpluses. At the same time, some Members of Congress have expressed concerns about the level of federal funding in light of the current federal fiscal condition. In addition, R&D funding decisions may be affected by differing perspectives on the appropriate role of the federal government in advancing science and technology.

As Congress undertakes the appropriations process it faces two overarching issues: (1) the direction in which the federal R&D investment will move in the context of increased pressure on discretionary spending and (2) how available funding will be prioritized and allocated. Low or negative growth in the overall R&D investment may require movement of resources across disciplines, programs, or agencies to address priorities. Congress will play a central role in defining the nation's R&D priorities as it makes decisions with respect to the size and distribution of aggregate, agency, and programmatic R&D funding.

America COMPETES Act Reauthorization

The America Creating Opportunities to Meaningfully Promote Excellence in Technology, Education, and Science (COMPETES) Act (P.L. 110-69) was enacted in 2007. The act, a response to concerns about U.S. competitiveness, authorized certain federal research, education, and innovation-related activities. In 2010, Congress passed the America COMPETES Reauthorization Act of 2010 (P.L. 111-358), extending and modifying certain provisions of the 2007 law, as well as establishing new provisions. Congressional appropriations have generally been below authorized levels, and the specific authorizations of appropriations in the 2010 act have expired. Following previous reauthorization efforts that inspired debate about such topics as the scientific peer review process, certain provisions of these acts were reauthorized and modified as part of the American Innovation and Competitiveness Act (AICA, P.L. 114-329), enacted at the end of the 114th Congress. The 115th Congress may consider additional provisions from the COMPETES acts that were not addressed through the AICA, such as expired authorizations of appropriations for the National Science Foundation (NSF) and the National Institute of Standards and Technology (NIST).

The COMPETES acts were originally enacted to address concerns that the United States could lose its advantage in scientific and technological innovation. Economists have asserted that economic, security, and social benefits accrue preferentially to nations that lead in scientific and technological advancement and commercialization. Some analysts have suggested that historical U.S. leadership in these areas is slipping, and in particular, some stakeholders have questioned the adequacy of federal funding for physical sciences and engineering research and the domestic production of scientists and engineers.

The COMPETES acts were designed to respond, in part, to these challenges by authorizing increased funding for the National Institute of Standards and Technology, National Science Foundation, and Department of Energy's Office of Science. Together, the acts also authorized certain federal STEM education activities, the Advanced Research Projects

Agency-Energy (ARPA-E), and prize competitions at federal agencies, among other provisions.

Those who have expressed opposition to aspects of the COMPETES acts have done so from several perspectives. Some critics question the existence of a STEM labor shortage and thus the need for programs aimed at increasing the number of STEM workers. Other critics agree with the assertion of a shortage, but question whether the federal government should address it, believing that the market will make the necessary corrections to meet the demand. With respect to U.S. competitiveness, some analysts prefer alternative approaches to those proposed in the COMPETES acts, such as research tax credits or reducing regulatory costs. Other analysts object to the financial cost associated with the COMPETES acts, given concern about the federal budget deficit and debt.

Regulations for Federally Funded Research at Academic Institutions

For decades, the federal government and academic research institutions have been partners in supporting American innovation, competitiveness, and economic growth. The federal government is the largest source of academic research and development (R&D) funding in the United States, providing funds through more than two dozen federal agencies, with the National Institutes of Health (NIH) and the National Science Foundation (NSF) providing the largest portions of federal R&D funding to U.S. colleges and universities.

As part of oversight of federal funding for academic research, Congress and federal agencies have established requirements through statutes, regulations, and guidance documents that U.S. universities and other research institutions must comply with when applying for, receiving, and reporting on the results of federal research grants. Such requirements seek to ensure transparency and effectiveness of federal funds, while helping to prevent waste, fraud, and abuse. Academic research institutions broadly recognize the need for federal regulations but have raised concerns about unintended consequences, such as reducing research productivity and the return on federal investments. Areas of concern frequently cited by researchers and academic administrators include the amount of time spent completing administrative tasks compared to conducting research; the increasing number, and lack of harmonization, of requirements across federal funding agencies; and the adequacy of stakeholder engagement in the review and modification of federal regulations.

Legislation was enacted in the 114th Congress that addressed some of the concerns, including the 21st Century Cures Act (P.L. 114-255), the American Innovation and Competitiveness Act (AICA, P.L. 114-329), and the National Defense Authorization Act for Fiscal Year 2017 (NDAA, P.L. 114-328). Enacted provisions addressed issues at specific agencies, including conflicts of interest disclosure, financial reporting, and subrecipient monitoring. Enacted provisions also addressed cross-agency efforts by directing the establishment of an advisory committee (Research Policy Board) with federal and non-federal stakeholders, as well as an interagency working group on federal research regulations.

Congressional oversight may be an important part of monitoring the progress of implementing the provisions enacted in the 114th Congress in a holistic way and evaluating their overall effectiveness. Further, Congress may broadly consider the appropriate balance between supporting the nation's academic research enterprise through efforts to streamline regulations and maintaining effective mechanisms for oversight, transparency, and accountability.

Technology Transfer from Federal Laboratories

Every year, approximately one-third of the federal government's R&D spending is obligated to federal laboratories, including federally funded research and development centers, in support of agency mission requirements. The technology and expertise generated by federal laboratories may have applications beyond the immediate goals or intent of the original R&D. Over the years, Congress has established various mechanisms—primarily through the Stevenson-Wydler Technology

Innovation Act of 1980 and subsequent legislation—to facilitate the transfer of technology and research generated from federal laboratories to the private sector where it can be further developed and commercialized.

Technology transfer from federal laboratories can occur in many forms. In some instances, it can occur through formal partnerships and joint research activities between federal laboratories and private firms, including through cooperative research and development agreements or CRADAs. In other cases, it can occur when the legal rights to a government-owned patent are licensed to a private firm.

Congressional interest in promoting the transfer of technology from federal laboratories is based on an interest in meeting social needs and promoting economic growth to enhance the nation's welfare and security. Economic benefits of a technology accrue when a product, process, or service is brought to the marketplace where it can be sold or used to increase productivity. In addition, cooperation with the private sector provides a means for federal scientists and engineers to obtain technical information from the private sector, which in some instances is more advanced than the federal government.

Despite the efforts of federal agencies and Congress to increase the effectiveness of technology transfer from federal laboratories to the private sector, the use of federal technologies has remained restrained. Critics argue that working with federal laboratories continues to be difficult and time-consuming. Proponents of current efforts assert that federal laboratories are open to interested parties, but it remains up to the private sector to use them. At issue is whether additional legislative initiatives and federal incentives are needed to encourage increased technology transfer from federal laboratories, or if the available resources are sufficient.

Adequacy of the U.S. Science and Engineering Workforce

The adequacy of the U.S. science and engineering (S&E) workforce has been an ongoing concern of Congress for more than 60 years. Scientists and engineers are widely believed to be essential to U.S. technological leadership, innovation, manufacturing, and services, and thus vital to U.S. economic strength, national defense, and other societal needs. Congress has enacted many programs to support the education and development of scientists and engineers. Congress has also undertaken broad efforts to improve science, technology, engineering, and math (STEM) skills to prepare a greater number of students to pursue S&E degrees. In addition, some policymakers have sought to increase the number of foreign scientists and engineers working in the United States through changes in visa and immigration policies.

Most experts agree that there is no authoritative definition of which occupations comprise the S&E workforce. Rather, the selection of occupations included in any particular analysis of the S&E workforce may vary depending on the objective of the analysis. The policy debate about the adequacy of the U.S. S&E workforce has focused largely on professional-level computer occupations, mathematical occupations, engineers, and physical scientists. Accordingly, much of the analytical focus has been on these occupations. However, some analyses may use a definition that includes some or all of these occupations, as well as life scientists, S&E managers, S&E technicians, social scientists, and related occupations. Among the key indicators used by labor economists to assess occupational labor shortages are employment growth, wage growth, and unemployment rates.

Many policymakers, business leaders, academicians, S&E professional society analysts, economists, and others hold differing views with respect to the adequacy of the S&E workforce and related policy issues. These issues include the question of the existence of a shortage of scientists and engineers in the United States, what the nature of any such shortage might be (e.g., too few people with S&E degrees, mismatches between skills and needs), and whether the federal government should undertake policy interventions or rely upon market forces to resolve any shortages in this labor market.

Science, Technology, Engineering, and Mathematics Education

The term "STEM education" refers to teaching and learning in the fields of science, technology, engineering, and mathematics. Policymakers have had an enduring interest in STEM education. Popular opinion generally holds that U.S. students perform poorly in STEM subjects—especially when compared to students in certain foreign education systems—but the data paint a complicated picture. Over time, U.S. students appear to have made gains in some areas but may be perceived as falling behind in others.

Previous estimates of the federal STEM education effort found a range of between 105 and 254 STEM education activities at 13 to 15 federal agencies. However, as tracked over time by the Obama Administration, the number of federal STEM education activities appears to have dropped by half, from 254 to 125, between FY2010 and FY2016. While total federal funding for STEM education stayed about the same during this period—close to $3 billion annually—the effects of these changes on the federal STEM education effort overall are unknown.

The national conversation about STEM education frequently develops from concerns about the U.S. science and engineering workforce. Some observers assert that the United States faces a shortage of STEM workers; others dispute this claim. Many proponents argue that a general increase in STEM abilities among the U.S. workforce could benefit the nation regardless. On the other hand, some scholars oppose the use of education policy to increase the supply of STEM workers, either because they perceive such policies as overemphasizing the economic outcomes of education at the expense of other values (e.g., personal development or citizenship) or because they perceive the labor market as the more efficient mechanism for dealing with these issues.

Opinions differ as well on the appropriate scope, scale, and emphasis of federal STEM education policy. Some observers prefer policies aimed at lifting the STEM achievement of all students— such as teacher or faculty professional development; or changes in curriculum, standards, or pedagogy. Others emphasize policies designed to meet specific needs—

such as scholarships for the "best and brightest," federal workforce training, or programs for underrepresented groups.

Tax Incentives for Technological Innovation

The 115th Congress may consider new federal policies to promote technological innovation. Among the concerns fueling this interest is what some view as inadequate growth in domestic high-paying jobs in a range of industries in recent years. Three pathways to accelerating growth in these jobs are (1) faster rates of entrepreneurial business formation, (2) increased business investment in domestic R&D, and (3) greater domestic production of products and services derived from that research.

One way Congress can influence the rate of high-wage job creation is tax incentives for investment in innovation. Under current federal tax law, three provisions directly affect entrepreneurial business formation and business investment in R&D: (1) an expensing allowance for research expenditures under Section 174 of the tax code, (2) a non-refundable tax credit for increases in research expenditures above a base amount under Section 41, and (3) a full exclusion for capital gains from the sale or exchange of qualified small business stock held by the original investor for five or more years under Section 1202.

The tax credit and expensing allowance encourage companies to invest more in qualified research than they otherwise would by lowering the cost of capital for that research and boosting cash flow. Some argue that the current credit's incentive effect is too weak to increase business R&D investment to optimal levels. In their view, the credit's design has certain flaws that lead to uneven and arbitrary incentive effects and make it difficult for small startup companies to use the full amount of the credit in years when they have net operating losses. The credit was permanently extended at the end of 2015, while the expensing allowance has been a permanent tax provision since 1954.

Under Section 1202, a non-corporate investor may exclude 100% of any capital gain on qualified small business stock acquired after September 27, 2010. The exclusion is intended to boost equity investment in small startup firms in designated industries (particularly manufacturing) by reducing the tax burden on the returns to that investment relative to after-tax returns on alternative investments.

Recent research indicates that young startup firms account for most net U.S. job growth over time, but that access to capital remains a significant barrier to the formation of such firms. It is unclear what share of U.S. high-paying jobs are accounted for by small firms, but small entrepreneurial firms that survive and grow into large, successful companies can be a prolific source of such jobs. Industries that intensively use intellectual property (IP) directly and indirectly (through supply chains) accounted for 30% of U.S. jobs in 2014, and workers in those industries had average weekly earnings that were 46% larger than the average weekly earnings of workers in non-IP intensive industries in the same year.

The 115th Congress may consider whether new initiatives are needed to increase the rate of growth in domestic high-paying jobs. One option that might be explored is the creation of a tax incentive known as a patent or innovation box. Such an incentive lowers the tax burden on the income earned from the commercial use of qualified intellectual property, such as trademarks or patents. Depending on its design, a patent box could give companies investing in innovation a stronger incentive to expand their investment in U.S. R&D and production activities. Potential drawbacks to such a subsidy include its budgetary cost and the difficulty of justifying it on economic grounds.

AGRICULTURE

The federal government supports billions of dollars of agricultural research annually. The 115th Congress is likely to face issues related to funding this research and specific issues arising from advances in agricultural biotechnology and the use of antibiotics in food-producing animals.

Agricultural Research

Public investment in agricultural research has been linked to productivity gains, and subsequently to increased agricultural and economic growth. The U.S. Department of Agriculture's (USDA's) Research, Education, and Economics (REE) mission area has the primary federal responsibility of advancing scientific knowledge for agriculture. USDA-funded research spans the biological, physical, and social sciences related broadly to agriculture, food, and natural resources.

USDA conducts its own research and administers federal funding to states and local partners primarily through formula funds and competitive grants. The outcomes are delivered through academic and applied research findings, statistical publications, cooperative extension, and higher education.

USDA's research program is funded with nearly $2.9 billion per year of discretionary funding and about $120 million of mandatory funding. Congress traditionally reauthorizes and revises agricultural research programs in the five-year farm bill.

Since the 2014 farm bill (P.L. 113-79) expires on September 30, 2018, the next farm bill is expected to be debated in the 115th Congress. Some agricultural research programs will face reauthorization. Various agricultural, academic, and other interests may pursue changes to the scope or focus of those programs or request additional funding.

Agricultural Biotechnology

The 115th Congress might address issues regarding bioengineered foods labeling, or foods containing bioengineered ingredients, regulatory changes governing the introduction of genetically engineered (GE) plants and animals into the environment, and recent technical innovations that could raise new regulatory issues for agricultural biotechnology.

Currently, labeling GE food products is not required. The 114th Congress approved a bill (P.L. 114-216) in June 2016 that will establish a "national bioengineered food disclosure standard."

The U.S. Department of Agriculture (USDA) has two years to implement the labeling law. Food manufacturers can adopt either text, a symbol, or an electronic/digital link for identifying bioengineered foods. The labeling bill includes foods made through conventional genetic engineering technology as well as newer techniques in the definition of bioengineered foods.

P.L. 114-216 also requires USDA to conduct a study within a year of enactment that identifies potential technological factors that could affect consumer access to bioengineered food disclosure through electronic or digital methods such as QR codes on food products read by smart phones. Concerns have been raised that such digital methods of disclosure could have differential impacts on those without cell phones (e.g., the elderly, low-income families) and those without access to high-speed broadband. The required study is to specifically address the availability of wireless or cellular networks, availability of landline telephones in stores, and particular factors that might affect small retailers and rural retailers as well as consumers.

The development over the past several years of new technologies to genetically engineer plants, in particular novel gene-editing technologies, has raised new regulatory issues. USDA currently regulates GE plants under the Plant Protection Act (PPA; 7 U.S.C. §770). However, USDA has stated that newer technologies may fall outside the purview of the PPA and thus the Department might have no regulatory jurisdiction over plants genetically engineered using these new technologies. This has raised important questions about how such genetically engineered plants are to be regulated as they are introduced. As genetically engineered plant varieties created by these techniques become more common, and as the public becomes more aware that these varieties are not regulated under the PPA, Congress might choose to revisit the 1986 framework that governs U.S. biotechnology regulation.

Antibiotic Use in Food Producing Animals

The use of medically important antibiotics in food producing animals may interest the 115th Congress. Past Congresses have introduced legislation (in the 114th Congress, e.g., H.R. 1552 and S. 621) to restrict the use of some antibiotics in animals. In 2014, the Obama Administration launched the Combating Antimicrobial Resistant Bacteria (CARB) initiative, a cross-departmental effort to preserve effective antibiotics for critical public and animal health needs.

The U.S. Food and Drug Administration (FDA) evaluates human and animal drugs for safety and effectiveness under the Federal Food, Drug, and Cosmetic Act (FFDCA, 21 U.S.C. 301 et seq.). FDA is concerned about public health effects from certain antibiotic uses in food animals.

According to FDA, foods of animal origin may carry pathogens that cause foodborne infections, and antibiotic use in animals that produce these foods may render the infections untreatable due to antibiotic resistance.

In response to concern about antibiotic resistance, FDA issued guidance on the judicious use of antibiotics in animals. FDA finalized Guidance for Industry #213 in December 2013, giving animal drug sponsors three years to withdraw antibiotics for production uses (i.e., growth promotion or feed efficiency) and to update evidence for treatment and preventive or control uses. Guidance #213 became effective January 1, 2017. After this date, medically important antibiotics may only be used when necessary to treat and prevent diseases.

In June 2015, FDA finalized rules (80 *Federal Register* 31708) for the Veterinary Feed Directive (VFD)—prescriptions for animal drugs used in feed—that builds on FDA policy on judicious antibiotic use. Since January 1, 2017, all medically important antibiotics used in feed require a VFD. This requires that producers have an established veterinarian-client-patient relationship (VCPR) as defined in state regulations, or federal VFD regulations for states without VCPR requirements. A valid VCPR requires licensed veterinarians to be familiar with clients' farm and animals. Producers will now need prescriptions for non-feed antibiotics that previously were available over-the-counter. Finally, in May 2016, FDA issued a final rule (81 *Federal Register* 29129) to expand its reporting on the distribution and use of antibiotics.

Related to antibiotic resistance efforts, at the end of 2014, the U.S. Department of Agriculture (USDA) released its Antimicrobial Resistance Action Plan, with initiatives to develop research and collect information on antibiotic use in animals. USDA also requested additional funds in FY2017 for its antibiotic resistance activities.

BIOMEDICAL RESEARCH AND DEVELOPMENT

Advances in science and technology related to biomedical research and development underpin improvements in human health and quality of life. Some of the biomedical R&D issues that the 115th Congress may face include those related to the budget and oversight of the National Institutes of Health, the role the Food and Drug Administration in approving new medicines and laboratory tests, and advances in precision medicine and genomic editing.

National Institutes of Health: Budget and the 21st Century Cures Act

The National Institutes of Health (NIH) is the lead federal agency conducting and supporting biomedical research. The agency's budget of about $31.5 billion funds basic, clinical, and translational research in NIH's laboratories as well as in research institutions nationwide. The extramural research program (83% of the NIH budget) provides grants, contracts, and training awards to support over 30,000 individuals at more than 2,500 universities, academic health centers, and research facilities. Over a five-year period, FY1999-FY2003, the NIH budget doubled in current dollars, but since that time constraints on discretionary spending have decreased budget growth below the rate of inflation. In constant dollars, NIH funding peaked in FY2003 (not counting FY2009 stimulus funding); FY2016 NIH funding was 19% below the FY2003 peak.

The 21st Century Cures Act (P.L. 114-255), signed by President Obama on December 13, 2016, authorizes $4.8 billion for NIH over a 10-year period (FY2017-FY2026), averaging slightly under a half billion per year, or a 1.5% increase per year for the agency. However, the funding is not guaranteed and must be appropriated each year in subsequent appropriations acts. In addition, the funds may only be used for four specified projects: (1) the Precision Medicine Initiative, $1.455 billion for FY2017-FY2026; (2) the Brain Research through Advancing Innovative Neurotechnologies (BRAIN) Initiative, $1.511 billion for FY2017-FY2026; (3) cancer research, $1.8 billion for FY2017- FY2023; and (4) regenerative medicine using adult stem cells, $30 million for FY2017-FY2020. The new law makes a number of other policy changes, such as modifying the NIH strategic planning process, altering the agency's reporting requirements, and reducing the administrative burden for researchers, among other things.

The Food and Drug Administration (FDA) and the 21st Century Cures Act

The Food and Drug Administration (FDA) regulates the safety of foods (including dietary supplements), cosmetics, and radiation-emitting products; the safety and effectiveness of drugs, biologics (e.g., vaccines), and medical devices; as well as public health aspects of tobacco products. In the 114th Congress, both the House and Senate introduced bipartisan legislation to support medical innovation, primarily through reforms to the NIH and changes to the drug, biologic, and device approval pathways at the FDA. On December 13, 2016, the 21st Century Cures Act (P.L. 114-255) was signed into law.

Among other things, Division A of the act contains provisions to accelerate development and review of certain medical products; for example, by directing the Secretary to facilitate an "efficient development program for, and expedite review of" eligible regenerative advanced therapies (including cell therapy, therapeutic tissue engineering products, human cell and tissue products); creating a limited population pathway for antibacterial drugs to treat a serious or life-threatening infection in a limited population of patients with unmet needs; and establishing a priority review program to incentivize development of treatments for agents that present a national security threat.

To help fund the activities and programs authorized in 21st Century Cures Act, the law creates an FDA Innovation Account, to which a total of $500 million is authorized to be transferred over a nine-year period (FY2017-FY2025). Release of funds from the FDA Account is controlled through the annual appropriations process. The 115th Congress will likely begin to see implementation of these programs and activities.

Oversight of Laboratory-Developed Tests (LDTs)

In vitro diagnostic (IVD) devices provide information that is used by clinicians and patients to make health care decisions. IVDs are used in laboratory analysis of human samples and include commercial test products and instruments used in testing, among other things. Laboratory-developed tests (LDTs) are a class of IVD that is manufactured, including being developed and validated, and offered, within a single laboratory. LDTs may sometimes be referred to as "home-brew tests." Genetic tests are a type of diagnostic test that analyzes various aspects of an individual's genetic material (DNA, RNA, chromosomes, and genes). Most genetic tests are LDTs.

The regulation of LDTs has been the subject of debate over the past decade. The Food and Drug Administration (FDA) has traditionally exercised enforcement discretion over LDT regulation. This means that most LDTs, and most genetic tests, have not been subject to FDA premarket review and therefore have not received FDA clearance or approval for marketing. Given the growing use of LDTs and genetic tests in clinical medicine, the FDA has in recent years revisited whether, and the extent to which, LDTs should be regulated.

On July 31, 2014, the FDA notified the Senate Committee on Health, Education, Labor and Pensions and the House Committee on Energy and Commerce that it would be issuing draft guidance on LDT regulation. On October 3, 2014, the agency published a notice in the *Federal Register* announcing the availability of the guidance documents and requesting comments within 120 days to ensure their consideration in the development of final guidance. However, on November 18, 2016, the FDA indicated that it would postpone finalization of the draft guidance until the new Administration is in place.

Precision Medicine Initiative (PMI)

On February 25, 2016, the White House hosted a Precision Medicine Initiative (PMI) Summit to mark the one year anniversary of the initiative's launch, first announced in the 2015 State of the Union address. The mission of the PMI is "[t]o enable a new era of medicine through research, technology, and policies that empower patients, researchers, and providers to work together toward development of individualized care." The PMI primarily involves three federal agencies — the National Institutes of Health (NIH), the Food and Drug Administration (FDA), and the Office of the National Coordinator for Health Information Technology (ONC)—although other federal agencies have collaborated on and contributed to the effort. Precision medicine, also called personalized medicine, involves providing health care to an individual patient based on their unique characteristics, such as a genetic profile.

NIH has awarded multiple grants to begin building an extensive biobank, develop health care provider organizations (HPOs), and develop recruitment strategies for a million-person national research cohort program, now called the All of Us Research Program; NIH expects to begin enrolling participants in FY2017. To ensure the opportunity for participation of underserved individuals, NIH and the Health Resources and Services Administration (HRSA) awarded funding for a pilot program that is to determine infrastructure needs for health centers to serve as HPOs. In addition, FDA has developed *precision FDA* to facilitate data sharing and validation of new genomic assays in precision medicine.

The 115th Congress may be interested in policies that support ongoing or modified investments in precision medicine. In FY2016, NIH received $200 million for the PMI: $70 million for the National Cancer Institute and $130 million for the Common Fund for the research cohort. FDA received $2.392 million, and ONC did not receive funding. The FY2017 President's budget requested a total of $309 million for the PMI: $4 million to FDA, $5 million to ONC, and the remaining $300 million to NIH. Pursuant to the authorizations of appropriations in the 21st Century Cures Act (P.L. 114-255), NIH received $40 million—in additional appropriations—for the PMI via the second FY2017 Continuing Resolution (P.L. 114-254), which provides funding through April 28, 2017.

CRISPR: Advanced Genome Editing

Researchers have long been searching for a reliable and simple way to make targeted changes to the genetic material of humans, animals, plants, and microorganisms. A new gene editing tool known as CRISPR offers the potential for substantial improvement over previous methods in terms of ease of use, ubiquity, and cost. Some scientists believe CRISPR could lead to advances across a broad range of areas—from medicine and public health to agriculture and the environment. Additionally, CRISPR could lead to the development of products previously viewed as not feasible (e.g., the development of pig organs for human transplant). Potential advances offered by CRISPR, however, have also raised concerns about the ethics of some applications; associated health and safety risks; and the adequacy of the current regulatory framework to mitigate these health and safety risks. Both the potential opportunities and risks associated with CRISPR may be of interest in the 115th Congress.

For example, CRISPR-related approaches are being considered by some researchers to reduce or eliminate Zika virus and malaria. Effective reduction or elimination of the mosquito that serves as the primary vector for the transmission of Zika or malaria could save lives and substantially reduce costs. However, a 2016 report from the National Academy of Sciences indicates that existing mechanisms may be inadequate to assess the potential immediate and long-term environmental and public health consequences associated with the use of the technology.

Additionally, in April 2015, and again more recently in April 2016, Chinese scientists published results of experiments that attempted to modify the genetic makeup of nonviable human embryos using CRISPR. While federal funds currently cannot be used for research involving human embryos, the Chinese study has sparked ethical debates about the use of the technology in human embryos or to make permanent heritable changes to the genome. Some researchers and others have called for the establishment of international norms and the harmonization of regulations for the use of genome editing technologies.

Microbial Pathogens in the Laboratory: Safety and Security

In addition to its general oversight of workplace safety, the federal government addresses the safety of laboratory personnel who work with infectious microorganisms through guidance such as Biosafety in Microbiological and Biomedical Laboratories (BMBL), published by the Department of Health and Human Services Centers for Disease Control and Prevention (CDC) and National Institutes of Health (NIH). BMBL sets "Biosafety Levels" for work with the highest-risk pathogens. BMBL guidance is often adopted as a requirement. For example, BMBL compliance is required of federal grant recipients.

Biosecurity requirements, to protect the public from intentional and unintentional releases of pathogens, were first mandated by Congress in 1996, and expanded through subsequent reauthorizations. The Federal Select Agent Program, administered jointly by CDC and the U.S.

Department of Agriculture (USDA) Animal and Plant Health Inspection Service (APHIS), oversees the possession of "select agents," certain biological pathogens and toxins with the potential to cause serious harm to public, animal, or plant health. All U.S. laboratory facilities— including those at government agencies, universities, research institutions, and commercial entities—that possess, use, or transfer select agents must register with the program and adhere to specified best practices. All persons given access to select agents must undergo background investigations conducted by the Federal Bureau of Investigation (FBI).

Several incidents involving the mishandling of select agents in federal laboratories occurred in recent years. Samples of decades-old but viable smallpox virus were found in an FDA laboratory on an NIH campus. Laboratories at CDC, one of the select agent regulatory agencies, had incidents involving the anthrax agent, a virulent avian influenza virus, and Ebola virus. A Department of Defense laboratory inadvertently distributed viable instead of killed samples of the anthrax organism. Each incident was attributed, at least in part, to lapses in protocol or some other form of human error. Several incident reports have recommended improvements in the "culture of safety" in laboratories, standardized microbial handling practices, and better incident reporting, among other measures.

The authorizations of appropriations for the Select Agent Program (7 U.S.C. 8401-8402 and 42 U.S.C. 262a) expired in 2007. Congress has continued to fund ongoing implementation and enforcement. The 115th Congress may revisit program authority or regulations in light of recent incidents to determine whether current safety and security measures are adequate, as well as whether they allow important research on these pathogens to proceed.

DEFENSE

Science and technology play an important role in national defense. The Department of Defense (DOD) relies on a robust research and development effort to develop new military systems and improve existing systems.

Issues that may come before the 115th Congress regarding the DOD's S&T activities include budgetary concerns and the effectiveness of programs to transition R&D results into fielded products.

Department of Defense Research and Development

The Department of Defense spends over $60 billion per year on research, development, testing, and evaluation (RDT&E). Roughly 80-85% of this is spent on the design, development, and testing of specific military systems. Examples of such systems include large integrated combat platforms such as aircraft carriers, fighter jets, and tanks, among others. They also include much smaller systems such as blast gauge sensors worn by individual soldiers. The other 15-20% of the RDT&E funding is spent on what is referred to as DOD's Science and Technology Program. The S&T Program includes activities ranging from basic science to demonstrations of new technologies in the field. The goal of DOD's RDT&E spending is to provide the knowledge and technological advances necessary to maintain U.S. military superiority.

DOD's RDT&E budget contains close to 1,000 individual line items. Congress provides oversight of the program, making adjustments to the amount of funding requested for any number of line items. These changes are based on considerations such as whether the department has adequately justified the expenditure or the need to accommodate larger budgetary adjustments.

RDT&E priorities and focus, including those of the S&T portion, do not change radically from year to year, though a few fundamental policy-related concerns regularly attract congressional attention. These include ensuring that S&T, particularly basic research, receives sufficient funding to support next generation capabilities; seeking ways to speed the transition of technology from the laboratory to the field; and ensuring an adequate supply of S&T personnel. Additionally, the impact of budgetary constraints, including continuing resolutions, on RDT&E may be of interest to the 115th Congress. Specifically, senior DOD officials have been describing the need to develop and implement a third offset strategy—a strategy aimed at identifying new and innovative ways to maintain the dominance of U.S. military capabilities into the future—which would likely require increased investment in RDT&E.

ENERGY

The science and technology related-energy issues that may come before the 115th Congress include those related to reprocessing spent nuclear fuel, advances in nuclear energy technology, and the development of biofuels and of ocean energy technology.

Reprocessing of Spent Nuclear Fuel

Spent fuel from commercial nuclear reactors contains most of the original uranium that was used to make the fuel, along with plutonium and highly radioactive lighter isotopes produced during reactor operations. A fundamental issue in nuclear policy is whether spent fuel should be "reprocessed" or "recycled" to extract plutonium and uranium for new reactor fuel, or directly disposed of without reprocessing. Proponents of nuclear power point out that spent fuel still contains substantial energy that reprocessing could recover. However, reprocessed plutonium can also be used in nuclear weapons, so critics of reprocessing contend that federal support for the technology could undermine U.S. nuclear weapons nonproliferation policies.

In the 1950s and 1960s, the federal government expected that all commercial spent fuel would be reprocessed, using "breeder reactors" that would convert uranium into enough plutonium to fuel additional commercial breeder reactors.

Increased concern about weapons proliferation in the 1970s and the slower-than-projected growth of nuclear power prompted President Carter to halt commercial reprocessing efforts in 1977, along with a federal demonstration breeder project. President Reagan restarted the breeder demonstration project, but Congress halted project funding in 1983 while continuing to fund breeder-related research and development by the Department of Energy (DOE). Under President Clinton, research on producing nuclear energy through reprocessing was largely halted, although some work on the technology continued for waste management purposes.

The George W. Bush Administration renewed federal support for reprocessing, proposing to complete a pilot plant by the early 2020s. The Obama Administration halted plans for the pilot plant and redirected DOE's Fuel Cycle Research and Development Program toward development of technology options for a wide range of nuclear fuel cycle approaches, including direct disposal of spent fuel (the "once through" cycle), deep borehole disposal, and partial and full recycling. The Obama Administration's FY2017 funding request for this program was $249.9 million, a 22.6% increase from the enacted FY2016 level of $203.8 million.

Another DOE program related to reprocessing policy is the Mixed Oxide Fuel Fabrication Facility (MFFF) under construction at the Department's Savannah River Site in South Carolina. MFFF would produce fuel for commercial nuclear reactors using surplus nuclear weapons plutonium, as part of an agreement with Russia to reduce nuclear weapons material. Critics of the project contend that MFFF would subvert U.S. nonproliferation efforts by encouraging the use of plutonium fuel. Because of rising costs, the Obama Administration proposed to halt the MFFF project in FY2017 and pursue alternative plutonium disposition options.

Advanced Nuclear Energy Technology

All currently operating commercial nuclear power plants in the United States are based on light water reactor (LWR) technology, in which ordinary water cools the reactor and acts as a neutron moderator to help sustain the nuclear chain reaction. DOE has long conducted research and development work on other, non-LWR nuclear technologies that could have advantages in safety, waste management, and cost. A growing number of private-sector firms are pursuing commercialization of advanced nuclear technologies as well.

Advanced nuclear energy technologies include high-temperature gas-cooled reactors, liquid metal-cooled reactors, and molten salt reactors (in which the nuclear fuel is dissolved in the coolant), among a wide range of other concepts. Research on advanced reactor coolants, materials, controls, and safety is carried out by DOE's Advanced Reactor Technologies program. The Obama Administration requested $73.5 million for the program in FY2017, a 27.7% reduction from the FY2016 level.

Private-sector nuclear technology companies contend that a major obstacle to commercializing advanced reactors is that NRC's licensing process is based on existing LWR technology. They have urged NRC to develop a licensing and regulatory framework that could apply to all nuclear concepts. They also have recommended a "staged review process" to provide conditional NRC approval for advanced reactor designs at key milestones toward the issuance of an operating license. NRC and DOE are currently implementing the Joint Advanced Non-Light Water Reactors Licensing Initiative to adapt existing general design criteria for LWRs for use by advanced reactor license applications.

Another proposal to promote advanced nuclear technology calls for DOE national laboratories to host reactor demonstration projects sponsored fully or partly by the private sector. Some public-private R&D on advanced nuclear technology is already being conducted at national labs under DOE's Gateway for Accelerated Innovation in Nuclear (GAIN) program. Legislation to promote private-sector reactor demonstrations at national labs and to require NRC to develop an advanced reactor licensing framework was considered in the 114th Congress.

Biofuels

Biofuels—liquid transportation fuels produced from biomass feedstock—are often described as an alternative to conventional fuels. Some see promise in producing liquid fuels from a domestic feedstock that may reduce dependence on foreign sources of oil, contribute to improving rural economies, and lower greenhouse gas emissions. Others regard biofuels as potentially causing more harm to the environment (e.g., air and water quality concerns), encouraging landowners to put more land into production, and being prohibitively expensive to produce. The debate about the feasibility of biofuels is complex, as policymakers consider a multitude of factors (e.g., feedstock costs, timeframe to reach substantial commercial-scale advanced biofuel production). The debate can be even more complicated when one considers that biofuels may be produced using numerous biomass feedstocks and conversion technologies. Thus, for each specific biofuel, a thorough assessment of the costs and benefits requires specific knowledge of the various factors involved.

Congress has expressed interest in biofuels for decades, with most of its attention on the production of "first-generation" biofuels (e.g., cornstarch ethanol). Farm bills have had a significant effect on biofuel research and development. Starting in 2002, the farm bills have contained an energy title with several programs focused on assisting biofuel production (see "Agriculture" section for additional farm bill related research). While commercial-scale production of "first-generation" biofuels is well established, commercial-scale production for some advanced biofuels (e.g., cellulosic ethanol) is in its infancy.

In 2007, Congress expanded one policy that has supported an increase in advanced biofuel production—the Renewable Fuel Standard (RFS). The RFS requires U.S. transportation fuel to contain a minimum volume of biofuel, a significant percentage of which is gradually to come from advanced biofuels. However, the RFS has been under scrutiny for various reasons, including the reduction by the Environmental Protection Agency (EPA) in the total renewable fuel volume below what was required by statute and concerns about RFS compliance. This has created significant uncertainty for certain stakeholders, with the result that some of the advanced biofuel targets are not being met. An overarching issue is that the policy may require more biofuel to be produced than can be used given the existing motor fuel distribution infrastructure and the limited fleet of passenger vehicles that are built to run on higher percentage blends of biofuels. A continuing issue is whether a domestic biofuel industry is necessary for national defense, and what, if any, role the military might take regarding biofuel production and purchase. Congress could also consider whether to modify various biofuel promotional efforts, or to maintain the status quo.

Off-Shore Energy Development Technologies

Technological innovations are key drivers of U.S. ocean energy development. They may facilitate exploration of previously inaccessible resources, provide cost efficiencies in a low-oil-price environment, address offshore safety and environmental concerns, and address obstacles in the emerging offshore renewable energy industry. Private industry, universities, and government are all involved in ocean energy R&D. At the federal level, the Department of Energy (DOE) and the Department of the Interior (DOI) both support ocean energy research.

The 115th Congress may consider issues related to deepwater oil and gas drilling technologies. Interest in expanding deepwater operations has prompted advances such as ultra-deepwater-capable drilling vessels and equipment. The oil and gas industry and federal regulators have also focused on safety improvements to reduce the likelihood of catastrophic oil spills. In April 2016, DOI released final safety regulations that tighten requirements for offshore blowout preventer systems and other well control equipment. Issues include the cost to industry of meeting the rule's technological requirements and the time required to do so.

Congress may also consider technology issues related to offshore drilling in the Arctic, where icy conditions and infrastructure gaps pose challenges for the economic viability and safety of mineral exploration. A focus of industry R&D is on technology to extend the Arctic drilling season beyond the brief periods where sea ice is absent—for example, by developing ice-capable mobile offshore drilling units (MODUs). DOI finalized safety regulations for Arctic exploratory drilling in July 2016. Some have argued that they are too costly for industry and give inadequate weight to available technologies (such as those for well capping) that could reduce safety costs. Others question whether any rules or technologies can adequately ensure drilling safety in the Arctic given the environmental risks.

Among renewable ocean energy technologies, only wind energy is poised for commercial application in U.S. waters. In December 2016, the first U.S. offshore wind farm, off of Rhode Island, began regular operations. Developers are exploring technologies to increase offshore turbine efficiency and reduce costs, including floating turbines for deep waters. Other research explores improvements to electrical infrastructure, such as integrating transmission networks for multiple projects. An issue for Congress is whether and how to support or incentivize R&D for wind and other ocean renewables.

ITER

ITER (formerly known as the International Thermonuclear Experimental Reactor) is an international fusion energy research facility currently under construction in Cadarache, France. When completed, ITER is to be the world's largest fusion reactor and the first capable of producing more energy than it consumes. Although the energy output from ITER will not be harnessed to produce electricity, fusion researchers see ITER as the next step toward implementation of fusion energy as a power source.

ITER is an international collaboration. Along with the United States, the partners are the European Union, China, India, Japan, Russia, and South Korea. The United States withdrew from the initial design phase of ITER in 1998 at congressional direction, largely because of concerns about cost and scope. The project was restructured, and the United States rejoined in 2003. The formal international agreement to build the facility was approved in 2006.

The European Union, as host, is responsible for 45% of the construction cost, while the United States and the other participating countries are responsible for 9% each. Most of the U.S. share (which was $145 million in FY2016) is being contributed in kind, in the form of components and equipment sourced mostly from U.S. companies, universities, and national laboratories.

In recent years, management issues, schedule delays, and cost growth have again made ITER controversial, with repeated proposals in Congress to terminate U.S. participation. A central concern is that U.S. funding for ITER may be crowding out funding for the domestic fusion research program. In 2016, the Department of Energy (DOE) submitted a congressionally mandated report that recommended continued U.S. participation through FY2018 and reevaluating U.S. participation prior to submittal of the FY2019 budget.

First operation of ITER is now planned for 2027. Once operational, the lifespan of the facility is expected to be approximately 20 years. During the operation phase, and during subsequent deactivation and decommissioning, the agreed U.S. cost share is 13%.

ENVIRONMENT AND NATURAL RESOURCES

Science and technology play an increasingly large role in environmental issues. Science- and technology-related environmental issues that may come before the 115th Congress include climate change science, carbon sequestration, water science, and desalination.

Climate Change Science and Technology

Climate change, including the policy questions of whether and how the federal government might address it, is likely to appear on the agenda of the 115th Congress. Science and technology considerations will underpin virtually all congressional deliberations on the topic. Despite portrayals in popular media about controversies in climate change science, almost all climate scientists agree on certain important points: the Earth's climate has been changing; human-related emissions of greenhouse gases (GHG) are accumulating in the atmosphere (as rising concentrations); and that rising concentrations will lead to additional global warming and other climate changes. A large majority of scientists also agree that most of the observed global climate change since the 1970s has been human-related, and that continued GHG emissions would lead to important adverse impacts—even catastrophic for some populations; some, however, consider that even under current GHG emission trajectories, the potential impacts would be modest and manageable. The focus of climate change science may shift increasingly toward greater precision in climate projections, impacts on society, and probabilistic characterization of uncertainties, especially to assist local and regional impact assessment and risk management decision making. In light of the growing consensus on the science of human-induced climate change, Congress may shift attention more toward assessment of options to manage risks. This may include the magnitude and emphasis of federal support for climate-related technology for GHG emission abatement, geoengineering, and adaptation or resilience to impacts.

U.S. Global Change Research Program (USGCRP) is an interagency mechanism, required by the Global Change Research Act of 1990 (P.L. 101-606) that coordinates and integrates global change research across 10 government agencies. For FY2017, USGCRP has expressed three thematic priorities: climate, water cycle extremes in the context of climate change, and methane cycling in the context of the carbon cycle.

Most climate change-related funding is aimed at advancing "clean energy." This is because most human-related GHG emissions come from production, distribution, and combustion of fossil fuels, particularly for electricity generation and transportation. Many analysts see a decades-long path to decarbonization of the world's energy economy as a primary option to halting the human influence on climate, while potentially providing security and health benefits; some see potential carbon capture and sequestration (CCS) technologies as key to preventing carbon dioxide emissions while preserving a large place for coal and other fossil fuels in the energy economy.

Members may deliberate on the appropriate degree and means of federal support for advancing and deploying new technologies. Because some innovative technologies are still in the research phase, and because of market inefficiencies and barriers to economical uptake of some technologies, Congress has provided billions of dollars of existing federal support, including numerous tax incentives, for energy technologies, unevenly distributed across types of technologies. The programs range from basic research, through technology development and demonstration of selected technologies, to incentives to promote their commercial deployment. Some focus on "supply-push" of technologies, while others emphasize "demand-pull." Existing programs support technologies that variously lead to greater GHG emissions (e.g., fossil fuel extraction and utilization technologies) or would lower GHG emissions (e.g., more efficient and renewable energy technologies). Cleaner energy technologies can produce public health benefits additional to climate benefits, while shifts in the energy economy can also pose transitional challenges to employment and communities. The magnitude of federal expenditures for climate change, their effectiveness, and priorities may be topics for Congress, particularly in light of budget objectives.

The 115th Congress may consider legislation that affects existing programs or establishes new ones. One priority voiced by many in Congress is tax reform; changes in tax incentives could influence the types and rates of technology advance. In addition, legislation has been proposed in past Congresses to levy GHG emission fees or "carbon taxes" to broaden the tax base, to reduce other distortionary taxes, and to promote cleaner technologies. More specifically, pricing of GHG emissions would promote demand for more efficient and lower-emitting technologies, potentially driving innovation and market-based technology change. Emission fees could also be used in part to finance basic research in which the private sector tends to underinvest because of its risks and the difficulty of capturing all the benefits of successful R&D.

Technologies to support resilience to future climate change have also been proposed in past Congresses. Because the climate will continue to change, due to both natural and human-related causes, Congress may address the federal role in facilitating effective private decision making to anticipate and be resilient to changes. It may also consider efforts already begun to incorporate climate change projections into agency management of federal resources, infrastructure and operations, and requirements and incentives in federal programs that may encourage or impede adaptation. Effective decisions would all depend on the adequacy and appropriate use of scientific information and available technologies.

Carbon Capture and Sequestration

Carbon capture and sequestration (or storage)—known as CCS—is a physical process that involves capturing manmade carbon dioxide (CO_2) at its source and storing it indefinitely before its release to the atmosphere. CCS could reduce the amount of CO_2 emitted to the atmosphere while allowing the continued use of fossil fuels at power plants and other large industrial facilities. An integrated CCS system would include three main steps: (1) capturing CO_2 at its source and separating it from other gases; (2) purifying, compressing, and transporting the captured CO_2 to the sequestration site; and (3) injecting the CO_2 into subsurface geological reservoirs. Following its injection into a subsurface reservoir, the CO_2 would need to be monitored for leakage and to verify that it remains in the target geological reservoir. Once injection operations cease, a responsible party would need to take title to the injected CO_2 and ensure that it stays underground in perpetuity.

The U.S. Department of Energy (DOE) has pursued research and development of aspects of the three main steps leading to an integrated CCS system since 1997. Congress has appropriated more than $7 billion in total since FY2008 for CCS research, development, and demonstration

(RD&D) at DOE's Office of Fossil Energy. Nearly half of total funding, $3.4 billion, came from the American Recovery and Reinvestment Act (P.L. 111-5). Authority to expend Recovery Act funding expired at the end of FY2015.

To date, no commercial ventures in the United States capture, transport, and inject large quantities of CO_2 (e.g., 1 million tons per year or more) solely for the purposes of carbon sequestration. However, the CCS RD&D program has embarked on commercial-scale demonstration projects for CO_2 capture, injection, and storage. The success of these demonstration projects will likely bear heavily on the future outlook for widespread deployment of CCS technologies as a strategy for preventing large quantities of CO_2 from reaching the atmosphere while power plants continue to burn fossil fuels, mainly coal. Congress may review the results from these demonstration projects as they progress in order to gauge whether DOE is on track to meet its goal of allowing for an advanced CCS technology portfolio to be ready by 2020 for large-scale demonstration and deployment in the United States.

The U.S. Environmental Protection Agency's (EPA's) final rule for reducing CO_2 emissions from new fossil fuel power plants, part of the Obama Administration's Clean Power Plan (CPP), found newly constructed power plants incorporating partial CCS to be the Best System of Emission Reduction (BSER). EPA determined that the BSER is technically feasible and available at reasonable cost. EPA based its claim, in part, on an example of demonstrated, full-scale operations in the electricity-generating industry, as well as on other smaller projects that are reasonably predictive of results at full scale. In a separate rule, also part of the CPP, EPA found that CCS was not the BSER for existing power plants.

Water Technologies, Science, and Infrastructure Research

Reliable water quantity and quality is essential for the U.S. population, ecosystems, and economy, including agriculture and energy production. Because of the diverse uses of water, federal research activities span numerous departments, agencies, and laboratories; the federal government also supports water research through grants to universities and other researchers. Recent droughts, flood disasters, and drinking water contaminations (e.g., algal toxins in Lake Erie and lead in drinking water in Flint, MI) have increased attention to water science and technology. The 115th Congress may consider multiple water research topics including:

- Water monitoring infrastructure and programs, including water quality monitoring activities, stream gages, and remote sensing investments (see "Earth-Observing Satellites");
- Water efficiency technologies and practices, such as improved irrigation technologies (see "Agricultural Research"), and science to support their adoption;
- Water augmentation technologies and science to support their adoption, including urban and agricultural groundwater recharge and stormwater capture techniques, water reuse technologies, and desalination (see "Desalination");
- Research on altering the operation of existing reservoirs and augmenting hydropower generation;
- Technologies and materials for monitoring and rehabilitating aging infrastructure;
- Access to water data (e.g., the Open Water Data Initiative); and
- Coordination and direction of the federal water science and research portfolio.

Various entities have produced reports discussing aspects of the water research portfolio. For example, a 2014 Government Accountability Office (GAO) report, *Freshwater: Supply Concerns Continue and Uncertainties Complicate Planning*, found that state water management activities could benefit from improved federal water data; and a 2016 GAO report, *Municipal Freshwater Scarcity: Using Technology to Improve Distribution System Efficiency and Tap Nontraditional Water Sources*, assessed utility-scale water technologies including leak detection, metering, wastewater reuse, stormwater capture, and desalination. The National Academy of

Sciences published reports on greywater and stormwater in 2016, wastewater produced from oil and gas development in 2016, Gulf of Mexico restoration monitoring in 2015, and wastewater reuse in 2012.

Desalination

Water shortages have heightened interest in desalination for augmenting water supplies. Desalination can create a new high-quality, local freshwater supply that is independent of weather conditions. At issue for Congress is what should be the federal role in supporting desalination adoption and desalination technology research. The federal government is currently involved in various desalination research activities that span multiple departments, including for military applications. Additionally, some states, universities, and private entities also undertake and support desalination research.

Desalination consists of the treatment of a feed water of seawater or less saline water known as brackish water. The treatment results in not only freshwater but also saline wastewater known as concentrate or brine. The most common desalination technology in the United States is reverse osmosis, which uses permeable membranes to separate freshwater from saline waters.

The U.S. prospects for desalinated water as a municipal water supply vary for brackish water and seawater. Some interior and coastal communities already are desalinating brackish sources to augment municipal water supplies; the United States is currently a global leader in brackish desalination adoption. Some coastal communities are looking to desalinate seawater or estuarine water; seawater desalination is more costly and requires more energy than brackish water desalination.

Although desalination costs have dropped in recent decades, significant further decline may not happen with existing technologies. For reverse osmosis, electricity expenses represent one-third to one-half of the operating cost. Researchers are pursuing emerging desalination technologies (e.g., forward osmosis and capacitive deionization) in an effort to improve desalination's energy efficiency (e.g., use of waste heat as an energy source), reduce its environmental impacts (e.g., less brine requiring disposal), and improve its cost competitiveness as an intermittent or base load water supply.

In the Water Infrastructure Improvements for the Nation Act (P.L. 114-322), the 114th Congress extended and expanded the authorization for the Department of the Interior to support desalination RD&D. In its FY2017 budget request, the Department of Energy proposed a new low-energy desalination manufacturing research initiative. Desalination issues before the 115th Congress may include how to prioritize federal desalination research, the appropriations level for federally supported desalination research and projects, and the evolving federal-state-local regulatory context for desalination projects.

HOMELAND SECURITY

The federal government spends billions of dollars supporting research and development to protect the homeland. Some of the issues that the 115th Congress may consider include how the Department of Homeland Security performs research and development and federal efforts to develop and procure new medical countermeasures against chemical, biological, radiological, and nuclear agents.

R&D in the Department of Homeland Security

The Department of Homeland Security (DHS) has identified five core missions: to prevent terrorism and enhance security, to secure and manage the borders, to enforce and administer immigration laws, to safeguard and secure cyberspace, and to ensure resilience to disasters. New technology resulting from research and development can contribute to all these goals. The Directorate of Science and Technology has primary responsibility for establishing, administering, and coordinating DHS R&D activities. The

Domestic Nuclear Detection Office (DNDO) is responsible for R&D relating to nuclear and radiological threats. Several other DHS components, including the Coast Guard, also fund R&D and R&D-related activities related to their missions.

Coordination of DHS R&D is a long-standing congressional concern. In 2012, GAO concluded that because so many components of the department are involved, it is difficult for DHS to oversee R&D department-wide. In January 2014, the joint explanatory statement for the Consolidated Appropriations Act, 2014 (P.L. 113-76) directed DHS to implement and report on new policies for R&D prioritization. It also directed DHS to review and implement policies and guidance for defining and overseeing R&D department-wide. In July 2014, GAO reported that DHS had updated its guidance to include a definition of R&D and was conducting R&D portfolio reviews across the department, but that it had not yet developed policy guidance for DHS-wide R&D oversight, coordination, and tracking. In December 2015, the explanatory statement for the Consolidated Appropriations Act, 2016 (P.L. 114-113) stated that DHS "lacks a mechanism for capturing and understanding research and development (R&D) activities conducted across DHS, as well as coordinating R&D to reflect departmental priorities." The Common Appropriations Structure that DHS introduced in February 2016 in its FY2017 budget request includes an account titled Research and Development for each DHS component. It remains to be seen whether this change will help to address congressional concerns about DHS-wide R&D coordination.

DHS has reorganized its R&D-related activities several times. DNDO and the Office of Health Affairs (OHA) were both created largely by reorganizing elements of the S&T Directorate. In the explanatory statement for the Consolidated and Further Continuing Appropriations Act, 2013 (P.L. 113-6), Congress directed DHS to evaluate the option of merging DNDO and OHA and realigning some of their functions, possibly including R&D, into other components. In 2015, DHS proposed the creation of a new Office of Chemical, Biological, Radiological, Nuclear, and Explosives Defense, made up of DNDO and OHA together with smaller elements of other programs. In 2016, it included this proposal in the FY2017 budget request. The 115th Congress may consider legislation to authorize the establishment and activities of this new office.

Chemical, Biological, Radiological, and Nuclear Medical Countermeasures

The anthrax attacks of 2001 highlighted the nation's vulnerability to biological terrorism. The federal government responded to these attacks by increasing efforts to protect civilians against chemical, biological, radiological, and nuclear (CBRN) terrorism. Effective medical countermeasures, such as drugs or vaccines, could reduce the impact of a CBRN attack. Policymakers identified a lack of such countermeasures as a challenge to responding to the CBRN threat. To address this gap, the federal government created several programs to encourage private sector development of new CBRN medical countermeasures. Despite these efforts, the federal government still lacks medical countermeasures for many CBRN threats, including Ebola.

The Biomedical Advanced Research and Development Authority (BARDA) and Project BioShield are two key pieces of the federal efforts supporting the development and procurement of new CBRN medical countermeasures. BARDA directly funds the advanced development of countermeasures through contracts with private sector developers. Project BioShield provides a procurement mechanism to remove market uncertainty for countermeasure developers. It allows the federal government to agree to buy a countermeasure up to 10 years before the product is likely to finish development. Recent Congresses have modified these and related programs to improve their performance, efficiency, and transparency to oversight. However, some key issues remain unresolved, including those related to appropriations, interagency coordination, and countermeasure prioritization. In addition to questions regarding the amount of funding, Congress may decide whether to return to funding Project BioShield through a multiyear advance appropriation.

Policymakers may consider whether the new planning and transparency requirements have sufficiently enhanced coordination of the multiagency countermeasure development enterprise. Additionally, Congress may consider whether the countermeasure prioritization process appropriately balances the need to address traditional threats such as anthrax and smallpox with the threat posed by emerging infectious diseases such as Ebola.

INFORMATION TECHNOLOGY

The rapid pace of advancements in information technology presents several issues for congressional policymakers, including those related to cybersecurity, broadband deployment, access to broadband networks and net neutrality, the Internet of Things, cryptography and law-enforcement challenges, and federal networking R&D.

Cybersecurity

The 113th and 114th Congresses saw significant legislative activity relating to cybersecurity, including the enactment of several laws with provisions on the security of federal information systems and the sharing of cybersecurity information across critical infrastructure sectors, among other issues. Those laws, along with more than 50 other statutes, presidential directives, and related authorities, provide a complex federal policy framework for U.S. cybersecurity.

The 115th Congress may face a number of significant issues related to cybersecurity, in addition to oversight of implementation of enacted laws. Among those issues are

- cybersecurity for critical infrastructure, given that most of it is owned by the private sector;

- prevention of and response to cybercrime, especially given its substantially international character;
- the relationship between cyberspace and national security; and
- ways that federal funding should be invested to protect information systems.

In addition to such short- and medium-term issues, Congress may consider responses to a number of long-term challenges, including the following:

- the degree to which information systems can be designed with security built in, in the face of economic obstacles and the other challenges;
- ways to correct an economic incentive structure for cybersecurity that has often been called distorted or even perverse, with cybercrime widely regarded as cheap, profitable, and comparatively safe for the criminals, while cybersecurity is often considered expensive and imperfect, with uncertain economic returns;
- finding consensus on a consistent and effective model for approaching cybersecurity, given stakeholders from different sectors and different work subcultures with varying needs, goals, and perspectives; and
- a rapidly evolving cyberspace environment that both complicates the threat environment and may pose opportunities for shaping the direction of that evolution toward greater security.

Broadband Deployment

Broadband—whether delivered wirelessly or via fiber, cable modem, or copper wire—is increasingly the technology underlying telecommunications services such as voice, video, and data. Since the initial deployment of broadband in the late 1990s, Congress has viewed broadband infrastructure deployment as a means to improve regional economic development, and in the long term, to create jobs. According to the Federal Communications Commission's (FCC's) National Broadband Plan, the lack of adequate broadband availability is most pressing in rural America, where the cost of serving large geographical areas, coupled with low population densities, often reduce economic incentives for telecommunications providers to invest in and maintain broadband infrastructure and service. The National Broadband Plan also identified broadband adoption as a problem, wherein one in three Americans have broadband available, but choose not to subscribe. Populations continuing to lag behind in broadband adoption include people with low incomes, seniors, minorities, the less-educated, non-family households, and the non-employed.

The 115th Congress may address a range of broadband-related issues. These may include the continued transition of the telephone-era Universal Service Fund from a voice to a broadband-based focus, infrastructure legislation that may include incentives for broadband buildout, reauthorization of the broadband loan program in the farm bill, the development of new wireless spectrum policies, and to what extent, if any, regulation is necessary to ensure an open Internet. Additionally, the 115th Congress may choose to examine the existing regulatory structure and consider possible revision of the 1996 Telecommunications Act and its underlying statute, the Communications Act of 1934. Both the convergence of telecommunications providers and markets and the transition to an Internet Protocol (IP) based network have, according to a growing number of policymakers, made it necessary to consider revising the current regulatory framework. How a possible revision might create additional incentives for investment in, deployment of, and subscribership to U.S. broadband infrastructure may be among many issues under consideration.

To the extent that Congress may consider various options for further enhancing broadband deployment, a key issue is how to develop and implement federal policies intended to increase the nation's broadband availability and adoption, while at the same time minimizing any deleterious effects that government intervention in the marketplace may have on competition and private sector investment.

Access to Broadband Networks and the Net Neutrality Debate

As policymakers continue to debate telecommunications reform, a major point of contention is whether action is needed to ensure unfettered access to the Internet. The move to place restrictions on the owners of the networks that compose and provide access to the Internet, to ensure equal access and non-discriminatory treatment, is referred to as "net neutrality." While there is no single accepted definition of "net neutrality," most agree that any such definition should include the general principles that owners of the networks that compose and provide access to the Internet (i.e., broadband access providers) should not control how consumers lawfully use that network, and should not be able to discriminate against content provider access to that network. A focal point in the debate centers on whether it is necessary for policymakers to take steps to ensure "unfettered" access to the Internet for content, services, and applications providers, as well as consumers, and if so, what these steps should be. Some policymakers contend that more specific regulatory guidelines are necessary to protect the marketplace from potential abuses which could threaten the net neutrality concept. Others contend that existing laws and policies are sufficient to deal with potential anti-competitive behavior and that additional regulations would have negative effects on the expansion and future development of the Internet.

What, if any, action should be taken to ensure "net neutrality" is part of the overall discussion regarding broadband regulation. As the marketplace for broadband continues to evolve, some contend that no new regulations are needed, and if enacted will slow deployment of and access to the Internet, as well as limit innovation. Others, however, contend that the consolidation of broadband providers, coupled with their diversification into content, has the potential to lead to discriminatory behaviors which conflict with net neutrality principles. The two potential behaviors most often cited are the network providers' ability to control access to and the pricing of broadband facilities, and the incentive to favor network-owned or affiliated content, thereby placing unaffiliated content providers at a competitive disadvantage.

A consensus on the net neutrality issue remains elusive. Some Members of Congress support the FCC's adoption of the 2015 Open Internet Order establishing regulations for broadband Internet access. Others, while acknowledging that some regulation may be needed, argue that the FCC has overstepped its authority and advocate that the FCC look to Congress for guidance to amend the current law to update FCC authority before action is taken. Still others argue that regulation of the Internet is not only unnecessary, but harmful. Broadband regulation and the FCC's authority to implement such regulations is an issue of growing importance in the wide ranging policy debate over broadband access.

The Internet of Things

The Internet of Things (IoT) may be a focal point of far-reaching debates during the 115th Congress. The term refers to networks of objects with two features—a unique identifier and Internet connectivity. Such "smart" objects can form systems that communicate among themselves, usually in concert with computers, allowing automated and remote control of many independent processes and potentially transforming them into integrated systems. Such objects may include vehicles, appliances, medical devices, electric grids, transportation infrastructure, manufacturing equipment, building systems, and so forth. The IoT may potentially impact homes and communities, factories and cities, and nearly every sector of the economy, both domestically and globally, among them agriculture (precision farming), health (medical devices), and transportation (self-driving automobiles and unmanned aerial vehicles).

An increasing number of these systems require access to radio frequency spectrum in order to connect to the Internet or other networks. The development of fifth-generation (5G) wireless technologies is likely to develop in tandem with the IoT.

Although the full extent and nature of impacts of the IoT remain uncertain, some economic analyses predict that it will contribute trillions of dollars to economic growth over the next decade. The IoT, for example, may be able to facilitate more integrated and functional infrastructure, especially in "smart cities," through improvements in transportation, utilities, and other municipal services. Sectors that may be particularly affected are agriculture, energy, government, health care, manufacturing, and transportation.

The federal government may play an important role in enabling the development and deployment of the IoT, including R&D, standards, regulation, and support for testbeds and demonstration projects. No single federal agency has overall responsibility for the IoT. Various agencies have relevant regulatory, sector-specific, and other mission-related responsibilities, such as the Departments of Commerce, Health, Energy, Transportation, and Defense, the National Science Foundation, the Federal Communications Commission, and the Federal Trade Commission.

The range of issues that might be the subject of congressional activity includes the following:

- security of objects and the systems and networks to which they are connected;
- privacy of the information gathered and transmitted by objects;
- standards for the IoT, especially with respect to connectivity;
- transition to a new Internet Protocol (IPv6) that can handle the anticipated exponential increase in the number of IP addresses required by the IoT;
- methods for updating the software used by IoT objects in response to security and other needs;
- energy management for IoT objects, especially those not connected to the electric grid; and
- the role of the federal government in development and deployment, standards, regulation, and communications, including the impact of federal rules regarding "net neutrality."

The Internet of Things represents more than devices connected through networks, and more than Internet or radio frequency spectrum policy. Its growth will likely require significant changes in—and coordination among—many government departments and agencies.

Cryptography and Law Enforcement "Going Dark"

Changing technology presents opportunities and challenges for U.S. law enforcement. Some technological advances have arguably opened a treasure trove of information for investigators and analysts. Others have presented unique hurdles. While some feel that law enforcement now has more information available to them than ever before, others contend that law enforcement is "going dark" as their investigative capabilities are outpaced by the speed of technological change. These hurdles for law enforcement include strong, end-to-end (or what law enforcement has sometimes called "warrant-proof") encryption; provider limits on data retention; bounds on companies' technological capabilities to provide specific data points to law enforcement; tools facilitating anonymity online; and a landscape of mixed wireless, cellular, and other networks through which individuals and information are constantly passing. As such, law enforcement cannot access certain information they otherwise may be authorized to obtain.

The tension between law enforcement capabilities and technological change has received congressional attention for several decades. For instance, in the 1990s the "crypto wars" pitted the government against technology companies, and this tension was highlighted by proposals to build in vulnerabilities, or back doors, to certain encrypted communications devices as well as to restrict the export of strong encryption code. In addition, Congress passed the Communications Assistance for Law Enforcement Act (CALEA; P.L. 103-414) in 1994 to help law enforcement maintain their ability to execute authorized electronic surveillance as telecommunications providers turned to digital and wireless technology.

The "going dark" debate originally focused on data in motion, or law enforcement's ability to intercept real-time communications. However, more recent technology changes have impacted law enforcement's capacity to access not only communications but stored content, or data at rest. The Obama Administration urged the technology community to develop a means to assist law enforcement in accessing encrypted data and took steps to bolster law enforcement's technology capabilities. In addition, policymakers have been evaluating whether legislation may be an appropriate response to the problems posed by encryption.

The Networking and Information Technology Research and Development Program

Congress passed the High-Performance Computing and Communications Program (HPCC) Act of 1991 (P.L. 102-194) to enhance the effectiveness of federally funded information technology (IT) R&D programs and to encourage coordination among agencies conducting such research.

Proponents of federal support of IT R&D assert that it has produced positive outcomes for the country and played a crucial role in supporting long-term research into fundamental aspects of computing. Such fundamentals may provide broad practical benefits, but generally take years to realize. Additionally, the unanticipated results of research are often as important as the anticipated results. Another aspect of government-funded IT research is that it often leads to open standards, something that many perceive as beneficial, encouraging deployment and further investment. Industry, on the other hand, is more inclined to invest in proprietary products and will diverge from a common standard when there is a potential competitive or financial advantage to do so. Supporters believe that the outcomes achieved through the various funding programs create a synergistic environment in which both fundamental and application-driven research are conducted, benefitting government, industry, academia, and the public. Critics, however, assert that the government, through its funding mechanisms, may be picking "winners and losers" in technological development, a role more properly residing with the market. For example, the size of the Networking and Information Technology Research and Development (NITRD) Program may encourage industry to follow the government's lead on research directions rather than selecting those directions itself.

The NITRD Program is funded through appropriations to its individual agencies, so support for it will likely be part of the federal budget debate in Congress.

PHYSICAL AND MATERIAL SCIENCES

Some of the policy issues in the physical and material sciences that the 115th Congress may address include funding and oversight of the National Science Foundation and the multiagency initiative supporting research and development in the emerging field of nanotechnology.

National Science Foundation

The National Science Foundation (NSF) supports basic research and education in the non-medical sciences and engineering and is a primary source of federal support for U.S. university research. It is also responsible for significant shares of the federal science, technology, engineering, and mathematics (STEM) education program portfolio and federal STEM student aid and support. Enacted funding for NSF in FY2016 was $7.463 billion.

The NSF's funding levels and congressional direction of funding have been long-standing issues of congressional concern. At various points in NSF's history, some policymakers have pursued a policy of authorizing large increases in the NSF budget over a defined period of time (e.g., a 100% increase over seven years, sometimes referred to as a "doubling path policy"). Actual appropriations have rarely reached authorized levels, and growth in NSF's budget has slowed in recent years. Advocates of large funding increases assert that steep and fast increases in NSF funding are necessary to ensure U.S. competitiveness. Other analysts argue that steady, reliable funding increases over longer periods of time would be less disruptive to the U.S. scientific and technological enterprise. Alternatively, some policymakers seek no additional increases in NSF funding in light of the federal deficit and spending caps. Additionally, some policymakers prefer to direct federal funding to research with a more applied or mission-oriented focus than that which is typically funded at NSF.

Policy issues of particular interest that the 115th Congress may continue to address include the selection, funding, and management of large-scale construction projects, scientific instruments, and facilities, including the use of management fees and the construction of new large research vessels; the foundation's grant-making process, including its peer review process; and the effectiveness and costs of NSF's use of non-federal personnel—through the Intergovernmental Personnel Act (IPA) program—often called "rotators." Further, analysts and legislators have periodically debated questions about prioritizing NSF funding for the physical sciences and engineering over funding for the social, behavioral, and economic sciences, as well as expanding support for multidisciplinary funding. Other lasting federal policy issues for the NSF focus on the balance between scientific independence and accountability to taxpayers; the geographic distribution of grants; NSF's role in broadening participation in STEM fields; support for various STEM education programs; and the production of data about the U.S. scientific and technological enterprise.

Nanotechnology and the National Nanotechnology Initiative

Nanoscale science, engineering, and technology—commonly referred to collectively as nanotechnology—is believed by many to offer extraordinary economic and societal benefits. Nanotechnology R&D is directed toward the understanding and control of matter at dimensions of roughly 1 to 100 nanometers (a nanometer is one-billionth of a meter). At this size, the properties of matter can differ in fundamental and potentially useful ways from the properties of individual atoms and molecules and of bulk matter.

Many current applications of nanotechnology are evolutionary in nature, offering incremental improvements in existing products and generally modest economic and societal benefits. For example, nanotechnology is being used in automobile bumpers, cargo beds, and step-assists to reduce weight, increase resistance to dents and scratches, and eliminate rust; in clothes to increase stain- and wrinkle-resistance; and in sporting goods to improve performance. Other nanotechnology innovations play a central role in current applications with substantial economic value. For example, nanotechnology is a fundamental enabling technology in nearly all semiconductors and is key to improvements in chip speed, size, weight, and energy use. Similarly, nanotechnology has substantially increased the storage density of non-volatile flash memory and computer hard drives. In the longer term, some believe that nanotechnology may deliver revolutionary advances with profound economic and societal implications, such as detection and treatment of cancer and other diseases; clean, inexpensive, renewable power through energy transformation, storage, and transmission technologies; affordable, scalable, and portable water filtration systems; self-healing materials; and high-density memory devices.

The development of this emerging field has been fostered by significant and sustained public investments in nanotechnology R&D. In 2001, President Clinton launched the multi-agency National Nanotechnology Initiative (NNI) to accelerate and focus nanotechnology R&D to achieve scientific breakthroughs and to enable the development of new materials, tools, and products. More than 60 nations subsequently established programs similar to the NNI.

Through FY2016, Congress has appropriated approximately $21.8 billion for nanotechnology R&D; the President requested $1.4 billion in FY2017 funding. In 2003, Congress enacted the 21st Century Nanotechnology Research and Development Act (P.L. 108-153), providing a legislative foundation for some of the activities of the NNI, establishing programs, assigning agency responsibilities, and setting authorization levels through FY2008. Legislation has been introduced in successive Congresses to amend and reauthorize the act though none has been enacted into law. Congress has directed its attention primarily to three topics that may affect the realization of nanotechnology's hoped-for potential: R&D funding; U.S. competitiveness; and environmental, health, and safety (EHS) concerns.

SPACE

Congress has historically had a strong interest in space policy issues. Space topics that may come before the 115th Congress include the funding and oversight of the National Aeronautics and Space Administration (NASA) and issues related to the commercialization of space and to Earth-observing satellites.

NASA

Spaceflight has attracted strong congressional interest since the establishment of NASA in 1958. Issues include the goals and strategy of NASA's human spaceflight program, the impact of constrained budgets on NASA's other missions, and the future of NASA's Earth Science program. The 115th Congress will also likely address NASA reauthorization legislation, which was considered in both chambers in the 113th and 114th Congresses but not enacted.

With the end of the space shuttle program in July 2011, the United States lost the capability to launch astronauts into space. Since that time, NASA has relied on Russian spacecraft for crew transport to the International Space Station (ISS). For ISS cargo transport, NASA-contracted U.S. commercial flights have been delivering payloads of supplies and equipment since October 2012.

As directed by the NASA Authorization Act of 2010 (P.L. 111-267), NASA is pursuing a two-track strategy for human spaceflight. First, for transport to low Earth orbit, including the ISS, NASA is supporting commercial development of a crew transport capability like the commercial cargo capability achieved in 2012. Commercial crew transportation services are scheduled to become operational in 2018. The Government Accountability Office has reported that this date may slip to 2019.

Second, for human exploration beyond Earth orbit, NASA is developing a new crew capsule called Orion and a new heavy-lift rocket called the Space Launch System (SLS). The first crewed test flight of Orion and the SLS is scheduled for 2023. Details of the subsequent exploration missions of Orion and the SLS remain to be determined.

Rapid developments in the commercial space sector may change the relationship between NASA and industry. For example, in early 2017, SpaceX announced plans for a commercial flight in 2018 that would carry two passengers around the Moon and back. Some observers see this sort of development as potentially competing with NASA's human spaceflight plans. More broadly, the emergence of new commercial capabilities in space may present NASA with new opportunities for private-public partnerships or may shift its R&D priorities.

The 2010 authorization act authorized funding increases for NASA that were not subsequently appropriated. In considering reauthorization, the 115th Congress may examine whether reduced budget expectations require corresponding changes to planned programs. One common concern is that the cost of planned human spaceflight activities may mean less funding for other NASA missions, such as unmanned science satellites, aeronautics research, and space technology development.

NASA's Earth Science program, in which climate research is a major focus, is of particular congressional interest. Some in Congress have argued that other NASA activities should have higher priority or that some or all of NASA's Earth Science responsibilities should be transferred to other agencies. Supporters counter that space-based Earth observations are an integral part of NASA's science mission.

Commercial Space

A survey by the Department of Commerce found that U.S. companies had $62.9 billion in space-related sales in 2012. While U.S. government purchases provided much of this market, about one quarter of sales were within the commercial sector.

Although the commercial space industry has existed for several decades, some observers have identified an emerging "new space" sector of relatively new companies focused on private spaceflight at low cost. One factor driving this trend is NASA's reliance on commercial providers for access to the ISS, but "new space" companies are also focused on other markets. These include the launch of national security satellites for the Department of Defense, the launch of commercial satellites for U.S. and foreign companies, and even space tourism.

The Federal Aviation Administration (FAA) licenses commercial space launch and reentry, including commercial spaceports. As part of the FAA licensing process, the federal government indemnifies launch providers against certain third-party liabilities. The U.S. Commercial Space Launch Competitiveness Act (P.L. 114-90) extended this indemnification policy (for the ninth time since 1988) through September 2025. The act also extended through September 2023 a statutory moratorium that restricts the FAA's authority to regulate the safety of crewed spaceflight. The status of human spaceflight safety regulations has been a focus of recent congressional interest because of NASA's plans for commercial crewed flights to the ISS.

Several other federal agencies are also involved in the commercial space industry. The National Oceanographic and Atmospheric Administration (NOAA) licenses commercial remote sensing satellites. The Federal Communications Commission licenses the use of radio frequencies by commercial satellites and assigns locations for satellites in geostationary orbits. The National Transportation Safety Board investigates certain spacecraft accidents. The Department of Commerce Office of Space Commerce supports and promotes U.S. space commerce. Oversight of export controls on most aspects of commercial satellites shifted from the Department of State to the Department of Commerce in 2014. The 115th Congress may address coordination or simplification of these multiple agency roles.

Earth-Observing Satellites

The constellation of Earth-observing satellites launched and operated by the United States government performs a wide range of observational and data collecting activities, such as measuring the change in mass of polar ice sheets, wind speeds over the oceans, land cover change, as well as the more familiar daily measurements of key atmospheric parameters that enable modern weather forecasts and storm prediction. Satellite observations of the Earth's oceans and land surface help with short-term seasonal forecasts of El Niño and La Niña conditions, which are valuable to U.S. agriculture and commodity interests; identification of the location and size of wildfires, which can assist firefighting crews and mitigation activities; as well as long-term observational data of the global climate, which are used in predictive models that help assess the degree and magnitude of current and future climate change.

Congress continues to be interested in the performance of NASA, NOAA, and the U.S. Geological Survey in building and operating U.S. Earth-observing satellites. Congress has been particularly interested in the agencies meeting budgets and time schedules so that critical space-based observations are not missed due to delays and cost overruns. Concerns have been raised in Congress about the possibility of a "data gap" in the polar-orbiting weather satellite coverage. A near-term data gap could occur if the currently operating polar-orbiting weather satellite, the Suomi National Polar-orbiting Partnership (Suomi-NPP), fails before its successor, the first Joint Polar Satellite System (JPSS-1), is launched and operational sometime in 2017. JPSS-1 is designed to provide daily measurements from polar orbit that inform weather forecasts and storm predictions. The Government Accountability Office (GAO) has reported that a polar-orbiting weather satellite data gap would result in less accurate and timely weather forecasts and warnings of extreme weather events, which could endanger lives, property, and critical infrastructure.

On November 19, 2016, the GOES-R (Geostationary Operational Environmental Satellite-R) weather satellite launched and was placed into orbit. Now renamed GOES-16, it is the newest and most advanced weather satellite with sensors that should help improve hurricane tracking and intensity forecasts, prediction and warning of severe weather events, and rainfall estimates that will lead to better flood warnings. GOES-16 also carries the first operational lightning mapper in geostationary orbit, and will better monitor space weather—perturbations to the Earth's magnetic field caused by intense bursts of energy from the sun. GOES-16 is the first of a series of satellites that are planned for geostationary orbit coverage through 2036.

INDEX

P

R

S

T

U

W